106
Advances in Biochemical Engineering/Biotechnology

Series Editor: T. Scheper

Advances in Biochemical Engineering/Biotechnology

Series Editor: T. Scheper

Recently Published and Forthcoming Volumes

Biosensing for the 21st Century
Volume Editors: Renneberg, R., Lisdat, F.
Vol. 109, 2008

Biofuels
Volume Editor: Olsson, L.
Vol. 108, 2007

Green Gene Technology
Research in an Area of Social Conflict
Volume Editors: Fiechter, A., Sautter, C.
Vol. 107, 2007

Cell Separation
Fundamentals, Analytical and Preparative Methods
Volume Editors: Kumar, A., Galaev, I.Y., Mattiasson, B.
Vol. 106, 2007

White Biotechnology
Volume Editors: Ulber, R., Sell, D.
Vol. 105, 2007

Analytics of Protein-DNA Interactions
Volume Editor: Seitz, H.
Vol. 104, 2007

Tissue Engineering II
Basics of Tissue Engineering and Tissue Applications
Volume Editors: Lee, K., Kaplan, D.
Vol. 103, 2007

Tissue Engineering I
Scaffold Systems for Tissue Engineering
Volume Editors: Lee, K., Kaplan, D.
Vol. 102, 2006

Cell Culture Engineering
Volume Editor: Hu, W.-S.
Vol. 101, 2006

Biotechnology for the Future
Volume Editor: Nielsen, J.
Vol. 100, 2005

Gene Therapy and Gene Delivery Systems
Volume Editors: Schaffer, D. V., Zhou, W.
Vol. 99, 2005

Sterile Filtration
Volume Editor: Jornitz, M. W.
Vol. 98, 2006

Marine Biotechnology II
Volume Editors: Le Gal, Y., Ulber, R.
Vol. 97, 2005

Marine Biotechnology I
Volume Editors: Le Gal, Y., Ulber, R.
Vol. 96, 2005

Microscopy Techniques
Volume Editor: Rietdorf, J.
Vol. 95, 2005

Regenerative Medicine II
Clinical and Preclinical Applications
Volume Editor: Yannas, I. V.
Vol. 94, 2005

Regenerative Medicine I
Theories, Models and Methods
Volume Editor: Yannas, I. V.
Vol. 93, 2005

Technology Transfer in Biotechnology
Volume Editor: Kragl, U.
Vol. 92, 2005

Recent Progress of Biochemical and Biomedical Engineering in Japan II
Volume Editor: Kobayashi, T.
Vol. 91, 2004

Cell Separation

Fundamentals, Analytical and Preparative Methods

Volume Editors:
Ashok Kumar · Igor Yu Galaev · Bo Mattiasson

With contributions by
T. Aarvak · M. Bonyhadi · J. M. S. Cabral · S. Camacho · S. Craig
M. B. Dainiak · G. van den Engh · I. Y. Galaev · J. Hubble
S. F. Ibrahim · M. Kamihira · A. Kumar · E. Lien · B. Mattiasson
A. A. Neurauter · L. Nøkleby · R. E. Nordon · F. M. Plieva · E. Ruud

Springer

Advances in Biochemical Engineering/Biotechnology reviews actual trends in modern biotechnology. Its aim is to cover all aspects of this interdisciplinary technology where knowledge, methods and expertise are required for chemistry, biochemistry, micro-biology, genetics, chemical engineering and computer science. Special volumes are dedicated to selected topics which focus on new biotechnological products and new processes for their synthesis and purification. They give the state-of-the-art of a topic in a comprehensive way thus being a valuable source for the next 3–5 years. It also discusses new discoveries and applications. Special volumes are edited by well known guest editors who invite reputed authors for the review articles in their volumes.
In references *Advances in Biochemical Engineering/Biotechnology* is abbeviated *Adv Biochem Engin/Biotechnol* and is cited as a journal.

Springer WWW home page: springer.com
Visit the ABE content at springerlink.com

ISBN 978-3-642-09454-5 e-ISBN 978-3-540-75263-9

ISSN 0724-6145

DOI 10.1007/978-3-540-75263-9

Springer is a part of Springer Science+Business Media

springer.com

Cover design: WMXDesign GmbH, Heidelberg

Printed on acid-free paper 02/3141 YL – 5 4 3 2 1 0

Series Editor

Prof. Dr. T. Scheper

Institute of Technical Chemistry
University of Hannover
Callinstraße 3
30167 Hannover, Germany
scheper@iftc.uni-hannover.de

Volume Editors

Dr. Ashok Kumar

Department of Biological Sciences
and Bioengineering
Indian Institute of Technology
Kanpur, 208016-Kanpur
India
ashokkum@iitk.ac.in

Dr. Igor Yu Galaev

Department of Biotechnology
Center for Chemistry
& Chemical Engineering
Lund University
Lund 22100
Sweden
igor.galaev@biotek.lu.se

Prof. Dr. Bo Mattiasson

Department of Biotechnology
Center for Chemistry
& Chemical Engineering
Lund University
Lund 22100
Sweden
bo.mattiasson@biotek.lu.se

Editorial Board

Honorary Editors

Advances in Biochemical Engineering/Biotechnology Also Available Electronically

For all customers who have a standing order to Advances in Biochemical Engineering/Biotechnology, we offer the electronic version via SpringerLink free of charge. Please contact your librarian who can receive a password or free access to the full articles by registering at:

springerlink.com

If you do not have a subscription, you can still view the tables of contents of the volumes and the abstract of each article by going to the SpringerLink Homepage, clicking on "Browse by Online Libraries", then "Chemical Sciences", and finally choose Advances in Biochemical Engineering/Biotechnology.

You will find information about the

- Editorial Board
- Aims and Scope
- Instructions for Authors
- Sample Contribution

at springer.com using the search function.

Attention all Users of the "Springer Handbook of Enzymes"

Information on this handbook can be found on the internet at springeronline.com

A complete list of all enzyme entries either as an alphabetical Name Index or as the EC-Number Index is available at the above mentioned URL. You can download and print them free of charge.

A complete list of all synonyms (more than 25,000 entries) used for the enzymes is available in print form (ISBN 3-540-41830-X).

Save 15%

We recommend a standing order for the series to ensure you automatically receive all volumes and all supplements and save 15% on the list price.

Preface

Cell separation, which was once limited to merely being a basic technique for fractionating different cell populations, has come a long way in the last two decades. New, advanced and more specific and selective techniques have emerged as the demand for isolating a specific cell type for various biological applications has increased. Efficient and cost-effective techniques for fractionation and isolation of target cell types are necessary to provide pure cell populations for diagnostics, biotechnological and biomedical applications. One can see a considerable need, both in biomedical research and in diagnostic medicine, for the specific separation of a discrete population of cells from a mixture. For example, in the field of tissue engineering, isolation of stem cells from tissues or organs is of particularly great importance. Moreover, understanding cell developmental pathways becomes increasingly significant as diagnosis and treatment of diseases turns more to the molecular level. The diagnosis of cell-related diseases requires methods of detection, isolation and the analysis of individual cells, regardless of their relative content in the tissue. Since cell-based therapies now turn towards more realistic medical options, developing an effective separation system for large-scale cell separation has become challenging research goal for cell biologists and biotechnologists. The ideal technique should provide in a short time a good yield of cells with high purity while maintaining cell function. Despite the growing need for methods to separate cells into cell subpopulations, the existing cell-separation techniques still have some limitations when the desired degree of performance on a preparative scale is required. We will see more research focus in this direction in the future.

The traditional techniques of microfiltration, ultrafiltration and ultracentrifugation, which exploit differences in cell size, shape and density, have remained the workhorses despite low specificity and problems with scaling up. Flow cytometry, where the target cells are labelled with an immunofluorescent probe, is an accurate and routinely used method. With new developments in flow cytometry, one may more commonly see the availability of such a robust technique at affordable costs and efficient timescales. On the other hand, magnetic bead separation technology has gained popularity within the fields of cell biology and medical microbiology and new products have appeared in the last decade that have simplified the cell-separation technology arena. Selective separation of different cell types is now possible by using specific antibody coated magnetic particles that interact with the cell surface markers. It is felt that despite expansion of techniques and an increase in the scientific and technological knowledge associated with it, few books have been dedicated to reviewing the current status of cell-separation technologies.

This special volume on cell separations discusses fundamental and applied aspects of the analytical and preparative cell-separation technologies. The aim is to enlighten the reader with the new developments in cell-separation technologies and at the same time provide sufficient knowledge with other existing and more commonly used techniques. The volume is comprised of contributions from subject experts from both academia and industry, focuses on the research and commercial aspects of cell-separation technology, and provides readers with broader choice. Unlike protein separation, the major challenge in cell separation has been the recovery of the cells in viable form after they are bound to the separation matrix, as cells bind more strongly through multipoint attachment. This is an important focus of the present work and one we believe will provide new insight to researchers in this field.

The first introductory chapter, *Methods in Cell Separations*, gives a brief overview of the most commonly used cell-separation methods, which are discussed in terms of cell fractionation yields, throughput and purity. The chapter gives a brief introduction to some of these technologies, which are further discussed in more detail in separate chapters. The most important aspect of the introductory chapter is the updated information regarding commercially available reagents and methods for cell separations and the companies producing them. The remaining chapters in the volume provide the reader with information on each of the methods described, ranging from the basic technical principle of a method to its applications. The advantages and limitations of each technology reviewed are discussed, providing the reader with a critical view of each method and its applications. The introduction of cryogels as new tool for cell separations will provide readers with new ideas of their applications and this will certainly help researchers with the design of their own cell-separation projects. In conclusion, we believe the volume brings to the attention of researchers at all levels the variety of methods available for separating viable populations of cells and addresses the needs and challenges for future research in this growing field.

Lund, July 2007

Ashok Kumar
Igor Yu Galaev
Bo Mattiasson

Contents

Adv Biochem Engin/Biotechnol (2007) 106: 1–18
DOI 10.1007/10_2007_069

Published online: 28 July 2007

Methods in Cell Separations

Maria B. Dainiak[1,2] · Ashok Kumar[1,2,3] · Igor Yu. Galaev[1] · Bo Mattiasson[1] (✉)

[1]Department of Biotechnology, Center for Chemistry and Chemical Engineering, Lund University, P.O. Box 124, 221 00 Lund, Sweden
Bo.Mattiasson@biotek.lu.se

[2]Protista Biotechnology AB, IDEON, 223 70 Lund, Sweden

[3]Department of Biological Sciences and Bioengineering, Indian Institute of Technology Kanpur, 208016 Kanpur, India

Abstract Research in the field of cell biology and biomedicine relies on technologies that fractionate cell populations and isolate rare cell types to high purity. A brief overview of methods and commercially available products currently used in cell separations is presented. Cell fractionation by size and density and highly selective affinity-based technologies such as affinity chromatography, fluorescence-activated cell sorting (FACS) and magnetic cell sorting are discussed in terms of throughput, yield, and purity.

Keywords Cell separation · Purity · Recovery · Throughput · Surface antigen · Antibodies · Stem cells

Cell separation technology is an important tool in cell biology, immunology, stem cell research, and cancer research, allowing for the analysis and subsequent cultivation of defined cell populations. There is considerable need for the specific isolation and characterization of intact fetal cells in maternal blood for non-invasive prenatal diagnosis, stem cells for cell-based therapy, for separation of immunocompetent from non-immunocompetent cells, and of malignant cells from normal cells [1–5]. For example, highly efficient removal of tumor cells is required for autologous bone marrow transplantation. Enrichment steps are also necessary for the detection of low numbers of tumor cells in blood for disease prognosis [6]. Isolation and assessment of the functionality of T lymphocytes are vital in determining the progression rate in human immunodeficiency virus (HIV) infection and acquired immunodeficiency syndrome (AIDS), transplant rejection, and autoimmune disease [7]. Positive selection of hematopoietic stem cells (HSC) is now an essential step in cellular therapy. Due to their unique self-renewal and differentiation capacity HSC offer great potential for the treatment of hematological disorders, inborn errors of metabolism, and immunodeficiency [8, 9]. There are several important reasons for using highly purified progenitor cells in transplantation. As mentioned above, in autologous bone marrow transplantation it is necessary to remove tumor cells from the graft product. In allogeneic transplantation, the rationale for using purified stem cells is the depletion of T and B lymphocytes to avoid graft-versus-host-disease (GvHD), a common complication of allogeneic bone marrow transplantation, in which functional immune cells in the transplanted marrow recognize the recipient as "foreign" and mount an immunologic attack.

In general, stem cells purified from either a donor or the patient hold great promise in tissue engineering and in gene therapy. In the latter case, transducing an isolated progenitor cell should give rise to an unlimited number of similarly "repaired" cells, potentially curing single-gene disorders [10]. Apart from biomedicine and immunology, another wide area of applications of cell separation techniques is the isolation and detection of pathogenic microorganisms and protozoan parasites in various samples in food, clinical and environmental microbiology, and parasitology [11–14].

Due to the great importance of the cell sorting applications described above, the development of separation technologies has been focused on methods for increasing the purity and recovery of rare cell types, such as stem cells, antigen-specific B and T cells, and rare circulating tumor cells [15–17]. It is a challenging task, because target cells may occur at frequencies below one per million [17]. For example, the frequency of B cells specific for any particular antigen is usually less than 1% [18]. The percentage of HSC differs in different sources: it is less than 0.5% of peripheral blood cells but may increase to 1–5% after mobilization, whereas bone marrow mononuclear cells contain approximately 1.5% of HSC [3].

In traditional cell isolation processes, pre-enrichment steps are usually carried out for "debulking" the sample (for example, whole blood) using the differences in cell density or size [19]. The removal of unwanted cell populations is followed by more selective methods that can yield higher numbers of purified target cells. It is important that the isolated cells retain viability and biological function after the separation process. Thus, cell isolation techniques should be as mild as possible and avoid the use of interfering substances and shear forces during the process. As any isolation technique, the performance of cell separation is typically characterized by three parameters: throughput, purity, and recovery. Throughput is how many cells can be separated in a certain time, purity is the fraction of the target cells in the sample collected after the separation, and recovery is the fraction of the target cells obtained after the separation as compared to initially available target cells in the sample. It is still often necessary to choose between high purity with poor recovery and good recovery with lower purity [16].

The existing cell separation methodologies can be classified into two main groups. The first is based on physical criteria like size, shape, and density differences and includes filtration and centrifugation techniques [20–22]. These methods are commonly used for debulking heterogeneous samples. The second group comprises affinity methods such as capture on affinity solid matrix (beads, plates, fibers) [23–26], fluorescence-activated cell sorting

(FACS) [27, 28] and magnetic cell sorting [29, 30], which are based upon biochemical cell surface characteristics and biophysical criteria (in FACS). The basic principles of these methods, some examples of their main applications with an emphasis on stem cell isolation, and a few examples of cell separation products will be briefly described and discussed in terms of yield, purity, and scalability.

Table 1 describes some of the commercially available methods and reagents that make use of either the physical criteria or the affinity recognition of cells. A technique of cell separation by size employs materials with controlled pore sizes within the defined range (for example, woven matrix of nylon filaments with pores size of 5–200 μm, CellMicroSieves by BioDesign of New York). It permits isolation of cells and organelles and removal of pieces of dissociated cells from tissues, permitting only single cells to pass through. Such materials can be used as prefilters and for the removal of cellular debris prior to cell preparation.

Table 1 Cell separation products available commercially

Separation material and description	Application	Cell type	Starting material	Company
CellMicro sieves, woven matrix of pure nylon fibers	Cell fractionation based on size	Human, mouse, rat	Dissociated tissue, media and cells from a bioreactor	Biodesign of NY www.biodesignofny.com
Percoll and Ficoll–Paque- Denisty gradient medium	Separates cells based on density	Any	Blood and tissue suspensions	Amersham Biosciences www.amershambio sciences.com
PrepaCyte. Negative selection of cells. Agglutinates cells via cell surface antigen recognition. Sediment undesired cells	High yields of T-cells, NK cells and CD34+ progenitor cells	Human	Heparinized peripheral blood, bone marrow or cord blood	BioErgonomics www.bioe.com
NIM continuous density gradient for separation of human mononuculeated cells and neutrophils. NIM2 modified ficoll hapaque step gradient for animal blood separations	Separates neutrophils and mononucleated cells from whole blood	Human or rat	Heparanized blood	Cardinal Associates www.cardinal-sf.com

Table 1 (continued)

Separation material and description	Application	Cell type	Starting material	Company
Cellect negative selection affinity chromatography. Desired cells in flow through	T cells and subsets enrichment and purification. Tropoblast purification	Human, mouse and rat	Cell suspension	Cytovax Biotechnologies www.cytovax.com
Nycoprep, Lymphoprep and OptiPrep density gradient medium	Separates cells based on density	Human and animal	Blood and cell suspension	Life Technologies/ Gibco BRL www.lifetech.com
Cellgrow Lymphocyte separation medium. Isoosmotic polysucrose media with low viscosity	Isolation of lymphocytes from diluted whole blood	Human	Heparanized blood	Mediatech www.cellgro.com
ACT-CES combines density gradient separation with antibody selection using NPS (non-porous beads) to increase density of labeled cells	Stem cell enrichment, debulking and pre-FACS cell purification. Tumor purging and T-cell depletion and isolation	Human	Whole blood or buffy coat	MicraScientific www.micrasci.com
Gel microdrop single cells. Encapsulated agarose matrix for FACS	Isolation of cells based on protein secretion or effects of cytotoxic agents	Mammalian, bacterial, fungal	Cell suspension	One cell systems www.onecell.com
IsoCell. Negative selection affinity chromatography	Isolation and enrichment of T-cells, CD8 or CD4 cells	Human and mouse	Peripheral blood, spleen, lymph, thymus or bone marrow	Pierce Chemical www.pierecenet.com
Immunoaffinity column kits that isolate cells through negative selection. Cells are incubated with an antibody cocktail and cells of interest are in flow through	Enriches for T-cells, subsets, depending upon antibody cocktail	Human, mouse and rat	Density gradient separated leukocytes	R&D systems www.rndsystems.com

Table 1 (continued)

Separation material and description	Application	Cell type	Starting material	Company
Centricoll. Density gradient media for separation of living cells	Separates cells based on density	Any	Blood and tissue suspension	Sigma-Aldrich www.sigma-aldrich.com
Immulan coated beads. T cell, B cell and T cell subpopulation kits.	Separates cells based on affinity chromatography	Human, mouse, cat, dog or monkey	Blood, lymphocytes	Biotecx Laboratories www.biotecx.com
Immunocolumns: rapid affinity chromatography tools for isolation/enrichment of cell populations	Separates and enriches T, B-cells, CD4, CD8, CD3	Human, rat, mouse	Lymphocyte preparation	Cedarlane laboratories www.cedarlanelabs.com
Cryogel-SepA: Protein A-supermacroporous cryogel monolithic affinity columns	Separates T, B-lymphocytes, CD34+ cells, cells that can be labeled with antibodies to surface receptor	Human, rat, mouse	Blood, lymphocyte preparation, cell suspension	Protista Biotechnology www.protista.se

Separation of bioparticles by centrifugation involves the use of a density gradient medium, a reagent solution that forms a density gradient over the desired range. In this method, pH and osmolality are easily adjustable allowing preservation of the integrity of cells. The sample (for example blood) is layered on the medium and centrifuged to separate low-density from high-density cells. Several of the separation methods to be discussed rely on density gradient pre-enrichment techniques. Widely used media are Percoll and Ficoll-based [22, 31]. The latter is a neutral, highly branched, high-mass, hydrophilic polysaccharide that dissolves readily in aqueous solutions. Ficoll–Paque is widely used for the isolation of mononuclear cells from bone marrow, peripheral blood, and cord blood [31]. Ficoll can also be used to separate Langerhans islets from pancreatic tissue. The separated islets can then be used for transplantation into patients with type I diabetes. Percoll consists of colloidal silica particles of 15–30 nm diameter coated with poly(vinylpyrrolidone). Due to its heterogeneity in particle size, sedimentation occurs at different rates, creating isometric gradients. Calibration of Percoll gradients can be simplified with density marker beads. The beads are of predetermined densities and are color-coded (www.amershambiosciences.com).

Among the most active areas of application of centrifugation on Percoll medium are purification for *in vitro* fertilization of sperm of both human and bovine origin, and the separation of subcellular particles and blood cells [22]. As mentioned above, density gradient separation techniques are commonly performed as a debulking step prior to isolation of rare cell subtypes by affinity fractionation techniques. Recognition of cells through surface antigens may be used for the agglutination of the undesired cells, which are removed by sedimentation, and the specific cell population is enriched by negative selection. PrepCyte of BioErgonomics uses negative selection for the enrichment and purification of human T-cells, NK cells, and rare cell populations such as CD34+ progenitor cells.

Affinity separation methods for isolation of rare cell populations most commonly are based on the use of antibodies against differentially expressed cell-surface antigens. Enrichment of HSC is based on the use of antibodies against surface CD34 and CD133 antigens. The CD34 antigen is a transmembrane glycoprotein that is expressed on human hematopoietic progenitor cells and most endothelial cells, but is not found on mature blood cells and is not expressed by most solid tumors [32]. Several techniques based on the recognition of CD34 antigen have been used to purify HSC from human bone marrow, umbilical cord, and peripheral blood (Table 2). In the autologous setting, the use of separation techniques employing antibodies against CD34 antigen allows for effective purging of malignant cells that do not co-express this antigen [33]. However, some malignancies, such as acute lymphoblastic leukemia, co-express CD34 [34]. In such cases, separation techniques are based on the recognition of the CD133 antigen, which is not co-expressed by the malignant cells. CD133, a 120 kDa transmembrane glycoprotein is a unique stem cell marker and, in contrast to the CD34 antigen, is not expressed by late progenitors [35]. Highly purified CD133+ mobilized peripheral stem cells have already been successfully used for autologous and

Table 2 Some examples of separation technologies used for isolation of CD34+ and CD133+ cells

Technology	Product and company	Antigen	Cell source	Yield[a] %	Purity[a] %	Refs.
Immunoselection by magnetic beads	Dynabeads, Invitrogen Dynal AS	CD34	PB BM CB	> 5	> 30	[59]
	Isolex, Nexell	CD34	BM PB	40 53	93 90	
		CD34[b]	PB	76	97	

Table 2 (continued)

Technology	Product and company	Antigen	Cell source	Yield[a] %	Purity[a] %	Refs.
	Isolex50, Baxter Healthcare	CD34	BM	70	–	[3]
		CD34	PB	44	92	[67]
		CD34	CB	76	–	[68]
Immunoselection by submicro-scopic colloid magnetic beads	MACS, Miltenyi Biotec	CD34	CB	52	60 41	[69, 70]
		CD34	PB	71	97	[37, 71]
				77	98	[67]
				56	97 < 50	[61]
		CD133	PB	69	94	[33]
				81	93	[37]
		CD133	CB	96	82	[30]
		CD133	BM	23	82	[72]
High-speed fluorescence-activated cell sorting	High-speed cell sorter, SyStemix	CD34 Thy1	PB	60	88	[73]
Immuno-adsorption columns	Ceprate SC, CellPro	CD34	PB	50	–	[74]
				53	62	[75]
				35	72	[43]
		CD34	BM	46	64	[76]
Antibody panning selection	CD34 Cellector flask, AIS	CD34	BM	74	60	[77]
				15	33	[59]

BM bone marrow, *PB* peripheral blood, *CB* cord blood
[a] median values, if not indicated otherwise
[b] simultaneous + / – selection of CD34+ cell selection and T cell depletion

allogeneic transplantation in acute lymphoid leukemia [36, 37]. Enrichment of antigen-specific B cells is an example of another purification system, which involves the use of antigens for isolation of target cells [16]. B cells express surface immunoglobulin (sIg) molecules as antigen binding receptors [38]. Each B cell expresses sIg with a single antigen binding specificity, and only a few thousand B cells express receptors that are specific for any given antigenic epitope [39].

Typically, antibody labels are covalently linked to either a molecule (in immunofluorescent technology, FACS), a particle (in immunomagnetic technology), or a support matrix (immunoadsorbents in chromatography or bottom of a vessel in panning).

Immunoadsorbents are used either for negative selection, in which unwanted cells adhere to the column and the target cells pass through unretained [40], or for positive selection in which target cells bind to affinity adsorbent and, after washing, cells are released and eluted in a purified form [41, 42]. An example of negative selection is the enrichment of human T cells by selective removal of B lymphocytes using IgG-treated glass beads in Cellect and Cellect Plus columns marketed by Cytovax Biotechnologies, or in IsoCell kits from Pierce Chemicals [19]. Enrichment of the target cells using the Cellect system occurs in three steps. The first step involves the reconstitution of polyclonal antibodies and glass bead matrix and packing the column. Then, the cell suspension, which does not necessarily need to be pre-enriched by density gradient centrifugation, is applied to the column at a predefined flow rate. Too high a flow rate may prevent adhesion of the target cells and too low a flow rate leads to non-specific binding of the unwanted cells. Buffer is added to wash out target cells at optimum flow rates. Using this method, a removal of greater than 95% of B cells has been reported [19].

An example of positive selection is an immunoaffinity system, which is based on avidin coupled to polyacrylamide beads to selectively bind cells labeled with biotin-conjugated antibodies [41]. The method is used in Ceprate SC stem cell concentration system by CellPro, which consists of a column containing beads linked to avidin and a mechanical stirring device incorporated into the column. The purification procedure for HSC consists of the following steps. The cells are incubated with biotinylated anti-CD34 monoclonal antibodies. After washing to remove an excess of non-bound antibodies, the CD34 IgG-labeled cells are loaded onto the filtering pre-column to reduce non-specific binding. Then, the cells are applied to Ceprate avidin column. Non-labeled cells pass through the column while antibody-labeled cells adhere because of the biotin label to the avidin beads and are recovered from the adsorbent by mechanical agitation. In more detail, the effect of the mechanical forces on the cell detachment is discussed in a separate chapter in this volume – *Affinity adsorption of cells to surfaces and strategies for cell detachment.*

The yields of CD34+ cells obtained after this procedure are in the range of 35–58% [43–45]. The low yields have been a common problem of cell isolation procedures using immunoadsorbents due to the difficulties in recovering bound cells without effecting cell viability and function [46, 47]. The difficulty arises from the polyvalent nature of interactions between multiple receptors on the cell surface and multiple ligands on the affinity matrix [48].

Recently, an efficient method of detachment of affinity-bound cells by mechanical compression of novel affinity adsorbents, monolithic macroporous hydrogels, so called cryogels, has been introduced to overcome this problem [49, 50]. Cryogels have been designed specifically for chromatography of biological nanoparticles and are characterized by large (10–100 μm) interconnected pores, porosity of 89–91%, and high elasticity (Fig. 1). Protein A-cryogel columns have been used for separation of IgG-labeled B cells from T lymphocytes [42] and for positive selection of antiCD34-labeled CD34+ hu-

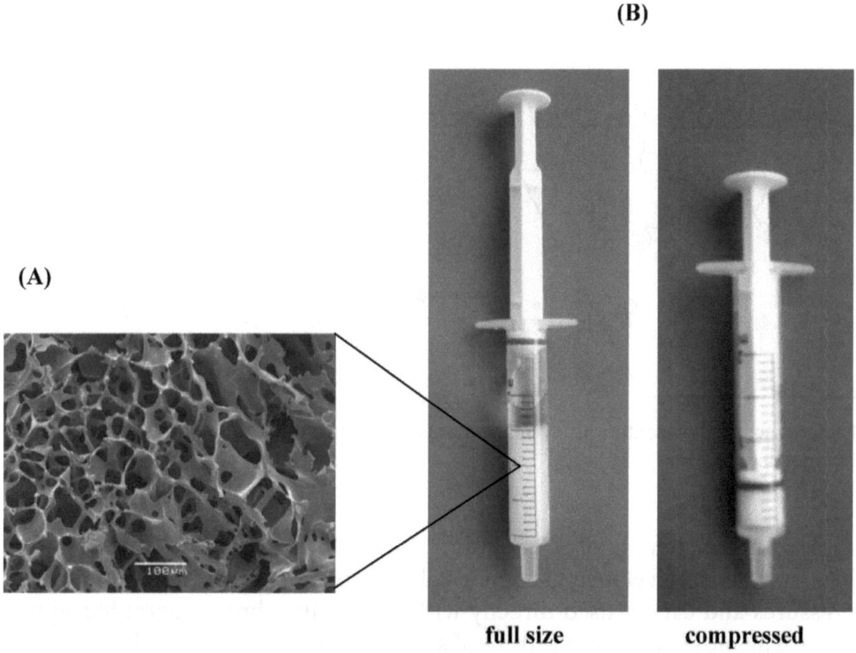

Fig. 1 Scanning electron micrograph of a cryogel (**A**). Full-sized and compressed protein A-cryogel (**B**)

man acute myeloid leukemia cells (KG-1) [49, 51]. Bound cells get released when the affinity matrix undergoes an elastic deformation (Fig. 2). This procedure does not require specific eluents, harsh elution conditions, or generation of shear forces. Up to 85% KG-1 cells with viability of 80–85% were recovered using this procedure [49]. This technique is discussed in more detail in another chapter in this volume – *Chromatography of living cells using supermacroporous hydrogels, cryogels.*

Monolithic matrices represent a novel generation of chromatographic supports, which are now being increasingly used for liquid and gas chromatography, capillary electrochromatography, and as supports for solid phase synthesis. The recent developments of monolithic adsorbents have shown tremendous potential in their application for the separation of proteins, oligonucleotides, plasmids, viruses, and nanoparticles.

Chromatography columns based on continuous beds (monolithic supports) offer advantages for separation of large biomolecules compared to traditional columns packed with beads or other particles. Because of superior mass-transfer and an open structure, monoliths provide very rapid separation and high resolution for large molecules. Earlier efforts at developing monolith supports have yielded some capabilities in analytical applications but have not translated to preparative use.

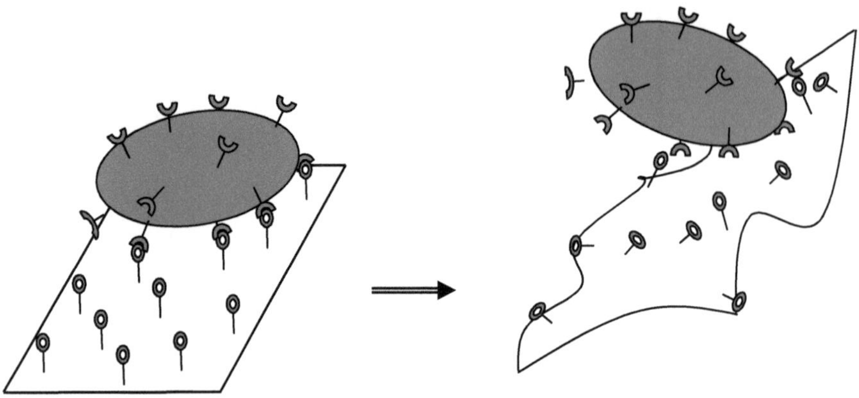

Fig. 2 Schematic presentation of the mechanism of detachment of bound cells induced by cryogel deformation

A complete new line of analytical and preparative monolithic columns have now been developed. The new columns operate under low to moderate pressures and can be used directly with all existing chromatographic equipment. The most important feature of such supports is that the mobile phase is forced to flow through the large pores of the medium. This enables rapid mass transfer, as well as high efficiencies, even at flow rates as high as 5–10 bed volumes per minute. In addition, the capacity and resolution are not affected by flow, therefore extremely fast separations are possible and the productivity of chromatographic processes can be increased by at least one order of magnitude over traditional chromatographic columns packed with porous particles.

Most of the porous monolithic matrices are produced from hydrophobic materials that show great promise in various biotechnological applications [52]. However, they have some limitations when dealing with separation processes of proteins, particles, cell organelles, or whole cells because of the non-specific interactions. In contrast, hydrophilic macroporous materials have a large application area in bioseparation allowing chromatographic separation of such particulate objects. The production of highly porous materials from hydrophilic polymers has not been addressed much so far. The production of macroporous agarose gels by double-emulsion procedures is one of the rare exceptions of successful production of hydrophilic macroporous material [53]. However, the technology suffers from poor reproducibility and the necessity of intensive washing of the macroporous material to remove the solvents and detergents used for double emulsion preparation. The local concentration of the polymer in the pore walls is the same as when non-macroporous agarose gel is formed. As hydrophilic polymers in aqueous environment bind a lot of water, the polymer phase in macroporous systems (walls of the macropores) has lower mechanical strength (rigidity) than polymeric phase of macroporous materials formed by hydrophobic polymers.

Thus, macroporous materials produced from hydrophilic polymers usually have poor mechanical characteristics and their application for chromatography is limited as these materials tend to collapse in chromatographic columns with increasing flow rates.

It is desirable to have a process producing hydrophilic macroporous material where the voids in the material are formed by "displacing" polymer chains from the pores into pore walls, so that the pore walls have high local polymer concentration as compared to the local polymer concentration in hydrophilic non-macroporous gels. The increased local polymer concentration in the pore walls will ensure sufficient mechanical stability of the material. Macroporous monolithic cryogels have been developed for cell affinity chromatography by carrying out polymerization under moderately frozen conditions.

An emerging technology utilizing immunoaffinity interactions for cell isolation is a high-speed fluorescence-activated cell sorting (FACS) [54–56]. FACS is a specific type of flow cytometry, which utilizes fluorescent affinity markers placed on the cells for the purpose of recognizing and sorting the cells. Flow cytometers measure relative fluorescence, size, and granularity of a single cell as it intersects a laser beam at a high velocity. In direct labeling of cells for FACS, a fluorochrome is chemically conjugated to a primary antibody that recognizes specifically an antigen of interest. In indirect staining, a fluorochrome-conjugated secondary antibody is used that recognizes the Fc region of a non-conjugated primary antibody. The three most common fluorochromes are fluorescein isothiocyanate (FITC), which emits green 530-nm light; phycoerythrin (PE), which emits yellow 578-nm light; and PE-Cy5 (PE conjugated to a cyanine dye), which emits light at 670 nm [54]. In advanced procedures, it is possible to label cells with two, three, or even four different fluorochromes simultaneously.

Figure 3 demonstrates the concept of labeling T cells in a heterogeneous suspension of cells containing T lymphocytes with FITC- and PE-conjugated antibodies that recognize and bind to CD8 and/or CD4 antigens on the cell surface, respectively. T cells expressing CD8 are labeled with the FITC-conjugated antibody and fluoresce green when excited by 488-nm light. T cells expressing CD4 fluoresce yellow, and CD4/CD8 double positive cells fluoresce both green and yellow. The principle and detailed description of flow cytometry is given in another chapter in this volume – *Flow cytometry and cell sorting*.

To isolate stem cells, SyStemix has developed a high-speed flow cytometer specially designed to perform the process under sterile conditions [57]. The process is based on the isolation of CD34+Thy1+ cell populations that contains multipotent T, B, and myeloid progenitor cells. Due to the very low frequency of CD34+Thy1+ cells in the starting population (Thy1 antigen is expressed on 5–25% of bone marrow CD34+ cells, and on 10–20% of CD34+ cells in mobilized peripheral blood [3]) several enrichment steps are necessary prior to flow cytometry. After removal of red blood cells, platelets, granulocytes, monocytes, and other differentiated myeloid cells in

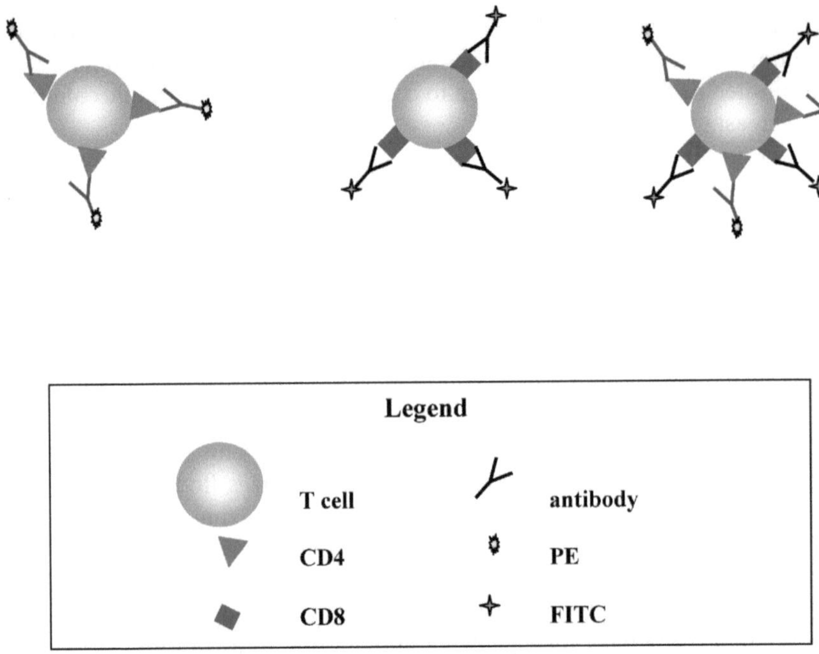

Fig. 3 Schematic presentation of fluorochrome-labeled antibodies binding to T cells that are either positive or negative for the surface antigens CD4 and CD8

the steps before cell sorting, the remaining cells are stained with directly conjugated anti-CD14 and anti-CD15 antibodies for the Lin+ population, and anti-CD34 and anti-Thy1 antibodies to identify the target population. A negative gate excludes Lin+ population, while a positive CD34+Thy1+ gate sorts the hematopoietic subset. A complete FACS procedure takes 2–3 h, while an overall processing time is 7–8 h [57].

Apart from purification of HSC, high-speed cell sorting is used for isolation of dendritic cells from peripheral blood, cell cycle analysis of rare populations, separation of X- and Y-chromosome bearing sperm for gender selection, and isolation of populations in different stages of T-cell development [58]. Despite high accuracy and the ability to purify extremely rare subpopulations, fluorescence-activated cell sorting has some major limitations, such as losses in the yield, the requirement for special technical expertise, and a comparatively low throughput, typically in the range between 10^4 and 10^5 cells per second [55]. Many experiments require far more cells than this. For instance, to perform proteomic analysis, about 10^7 cells would be needed to detect a protein expressed at roughly 1000 copies per cell, corresponding to a total starting population of approximately 10^8 cells [55]. Hence the process time would be up to 10^4 s, i.e., about 3 h. Running the high-speed sorter for many hours is costly and may pose problems in keeping the cells in their best condition during a long processing time. It is also noteworthy that FACS is one of the most expensive cell-sorting techniques with respect to initial capital investment and dedication of personnel [59].

Magnetic cell sorting is another widely used methodology for cell separation [3, 29]. The needs of separation technology are changing as the market changes. "In the market in cell separation applied to immunology one sees a real shift away from basic immunology toward more clinical work and applied immunology", which leads to a demand for new grades of products. Magnetic particles present one intriguing way of sorting cells. The separation process for the purification of target cells using magnetic labels and magnetic separators usually consists of the following steps. Cells are incubated with antibodies covalently linked to magnetic beads. Cell suspension containing magnetically labeled cells is placed in a magnetic field. Non-labeled cells flow through the magnetic separator, while labeled cells are retained. By removing the column from the magnet, the labeled fraction can be obtained. For selected applications the magnetic label has to be removed [29].

As in FACS technology, direct or indirect labeling of cells can be performed (Fig. 4). In direct methods, antibodies coupled to magnetic particles are added directly to cell suspension. Indirect labeling is usually performed when no direct antibodies coupled to magnetic beads are available. In that case, cells are first labeled either simply with a primary antibody or with a primary antibody that is for instance biotinylated, or fluorochrome-conjugated. In a second step, magnetic labeling is performed by using magnetic beads with coupled anti-immunoglobulin, anti-Biotin, or anti-Fluorochrome antibodies. Staining with fluorochrome allows for subsequent analysis of the separated cells by flow cytometry or microscopy.

Enrichment of target cells by magnetic separation techniques can be achieved using positive selection or depletion strategies. In depletion methods, the unwanted cells are magnetically labeled, get retained in the magnetic separator, and are eliminated from the cell mixture. The target cells pass through the separator and are collected as the enriched, unlabeled fraction. This method is usually carried out for the removal of unwanted cells, or in the cases when no specific antibody is available for target cells or labeling of the target cells is not desirable. However, if target cells are present in very low concentration, depletion methods may give low yield and purity due to non-specific loss of the cells to be isolated, or due to an insufficient removal of unwanted cells [29].

Positive selection is the most direct and specific way to isolate the target cells from a heterogeneous cell suspension. The target cells are magnetically labeled (directly or indirectly), get retained in a magnetic separator and finally are eluted after removal of the magnet as the enriched fraction. The drawback of this strategy is that an additional step, i.e., removal of the magnetic label, may be needed depending on the type of cells, their further application and type of magnetic particles.

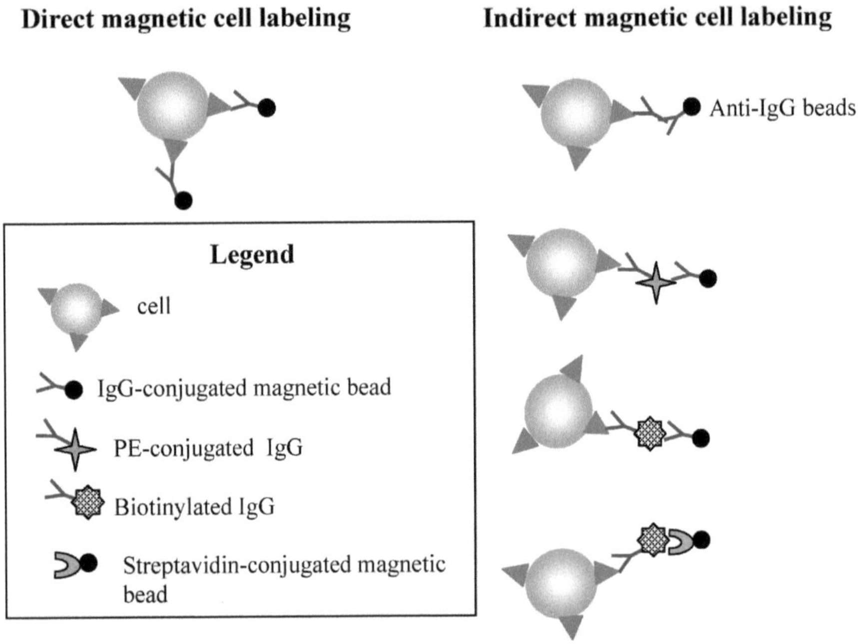

Fig. 4 Schematic presentation of direct and indirect magnetic cell labeling

Table 3 shows some examples of commercially available magnetic particles. Dynal's products which include magnetic particles for manipulating and stimulating T cells are described in the chapter *Cell isolation and expansion using Dynabeads* in this volume. Miltenyi Biotech, Polysciences, and other companies have developed cell separation products based on magnetic microbeads. Larger particles usually have to be removed, while cells labeled with submicroscopic magnetic beads (e.g., with MACS microbeads by Miltenyi Biotec, which are approximately 50 nm in size) may be used directly [3, 29, 60]. Besides, due to their composition of iron oxide and polysaccharide, the microbeads are biodegradable and typically disappear after a few days when the cells are cultured. They have been widely used for the isolation of CD34+ and CD133+ cells from different sources (Table 2) on both small- and clinical scales, and recently have received European Community approval for clinical use in Europe [30]. However, the purity and recovery in MACS typically have large variances [59]. For instance, a purity of less than 50% for CD34+ cells isolated using the Direct CD34+ Progenitor Cell Isolation Kit was reported by Kekarainen et al. [61]. The authors suggested an optimized scheme for the separation consisting of two-column method and an additional labeling step, which in fact means the performance of the same purification procedure twice, implying increased time and costs. In another CD34+ cell selection procedure based on the use of Isolex magnetic beads

Table 3 Some examples of commercially available magnetic particles

Name	Diameter (μm)	Composition	Immobilized compounds	Manufacturer
Isolex super-magnetic beads	4.5	Polystyrene	Secondary Abs	Baxter Healthcare
Dynabeads M-280 Dynabeads M-450 Dynabeads M-500	2.8 4.5 5	Polystyrene	Secondary Abs, anti-CD Abs, streptavidin	Invitrogen Dynal AS
Magnetic particles	1	Polystyrene	Streptavidin	Boehringer
Magnetic microparticles	1–2	Polystyrene	Protein A	Polysciences
MagaBeads	3.2	Polystyrene	Secondary Abs, streptavidin, protein A, protein G	Cortex Biochem.
SPHEROmagnetic particles	1–4.5	Polystyrene	Secondary Abs, streptavidin, biotin	Spherotech
XM200 microsphere	3.5	Polystyrene	Secondary Abs, protein A	Advanced Biotechnologies
MACS microbeads	0.05	Dextran	Secondary Abs, anti-CD Abs, streptavidin, biotin, anti-FITC, anti-PE	Miltenyi Biotec
Magnetic nanoparticles	0.09–0.6	Starch, dextran, chitosan	Streptavidin, protein A, biotin	Micro-caps
Ferrofluids	0.135, 0.175	Modified hydrophilic protein	Secondary Abs, streptavidin, protein A	Immunicon

and a fully automated device, Isolex300I, a combination of two methods, simultaneous positive selection of CD34 expressing cells and T cell depletion, resulted in 97% purity of the final HSC preparation [62].

Sometimes a combination of both methods, magnetic cell separation and FACS, is applied in order to achieve the desired purity in the preparation of target cells. For instance, in order to obtain highly purified HSC for high-dose therapy in cancer patients, CD34+ cell enrichment was carried out using Isolex immunomagnetic CD34+ positive selection followed by sorting using SyStemix's fluorescence-activated high-speed clinical cell sorter, resulting in

95.3% median purity of the stem cell product [63]. However, as mentioned earlier, despite the capability of generating pure population of cells, FACS is not suitable when isolation of large numbers of rare cells is required. Finally, it is also worth mentioning here that both separation procedures modify the cell membrane, thus neither technique is preferable for subsequent analysis or re-cultivation of the sorted cells [64].

It is rather obvious that commercial applications of the techniques discussed above have distinct advantages and disadvantages with respect to throughput, purity, and recovery in cell separation. As the demands placed on cell sorting technologies continue to increase, novel solutions to overcome the limitations of inherent coupling among these three competing performance parameters have to be found. Development of new matrices with properties different from those of conventional chromatographic beads used for protein chromatography is one of the possible solutions for overcoming such major obstacle as the problem of cell recovery in cell affinity chromatography [49]. As an alternative to the existing three main categories (immunofluorescent and immunomagnetic cell sorting and adsorption to immunomatrix) of cell separation technologies, novel approaches such as dielectrophoresis-activated cell sorting (DACS) are also being evaluated [65]. In DACS, mixtures containing rare cells labeled with particles that differ in polarization response are interrogated in the dielectrophoresis-activated cell sorter in a continuous-flow manner, wherein the electric fields are engineered to achieve separation between the dielectrophoretically labeled and unlabeled cells.

References

1. Geifman-Holtzman O, Makhlouf F, Kaufman L, Gonchoroff NJ, Hlotzman EJ (2000) Am J Obstet Gynecol 183:462
2. De Wynter EW, Coutinho LH, Pei X, Marsh JCW, Hows J, Luft T, Testa NG (1995) Stem Cells 13:524
3. Beaujean F (1997) Transfus Sci 18:251
4. Siewert C, Herber M, Hunzelmann N, Fodstad O, Miltenyi S, Assenmacher M, Schmitz J (2001) Recent Results Cancer Res 158:51
5. Racila E, Euhus D, Weiss AJ, Rao C, McConnell J, Terstappen L, Uhr J (1998) Proc Natl Acad Sci USA 95:4589
6. Prestvik WS, Berge A, Mork PC, Stenstad PM, Ugelstad J (1997) Preparation and application of monosized magnetic particles in selective cell separation. In: Häfeli U, Schutt W, Teller J, Zborowski M (eds) Scientific and clinical applications of magnetic carriers. Plenum, New York, London
7. Collins DP, Luebering BJ, Shaut DM (1998) Cytometry 33:249
8. Peters C, Steward CG (2003) Bone Marrow Transplant 31:229
9. Bhattacharya A, Slatter MA, Chapman CE, Barge D, Jackson A, Flood TJ, Abinun M, Cant AJ, Gennery AR (2005) Bone Marrow Transplant 36:295
10. Chan J, Siatskas C, Field J, Toh B-H, Alderuccio F (2006) Drug Discov Today: Disease Mechanisms 3:219

11. Islam D, Lindberg AA (1992) J Clin Microbiol 30:2801
12. Payne MJ, Campbell S, Patchett RA, Kroll RG (1992) J Appl Bacteriol 73:41
13. Deng MQ, Cliver DO, Mariam TW (1997) Appl Environ Microbiol 63:3134
14. Seesod N, Nopparat P, Hedrum A, Holder A, Thaithong S, Uhlen M, Lundeberg J (1997) Am J Trop Med Hyg 56:322
15. Shizuru JA, Negrin RS, Weissman IL (2005) Annu Rev Med 56:509
16. Kodituwakku AP, Jessup C, Zola H, Roberton AM (2003) Immunol Cell Biol 81:163
17. Gross HJ, Verwer B, Houck D, Hoffman RA, Recktenwald D (1995) Proc Natl Acad Sci USA 92:537
18. Oshiba A, Renz H, Yata J, Gelfand EW (1994) Clin Immunol Immunopathol 72:342
19. Vettese-dadey M (1999) The Scientist 13:21
20. Pretlow TG, Pretlow TP (eds) (1987) Cell separation - methods and selective applications. Academic, New York
21. Chianea T, Assidjo NE, Cardot PJP (2000) Talanta 51:835
22. Pertoft H (2000) J Biochem Biophys Method 44:1
23. Wigzell H, Anderson B (1968) J Exp Med 129:23
24. Truffa-Bachi P, Wofsy L (1970) Proc Natl Acad Sci USA 66:685
25. Edelman GM, Rutishauser U, Millette GF (1971) Proc Natl Acad Sci USA 68:2153
26. Wildt RMD, Steenbakkers PG, Pennings AH, Hoogen FHVD, Venrooij WJV, Hoet RM (1997) J Immunol Methods 207:61
27. Herzenberg LA, Parks D, Sahaf B, Perez O, Roederer M, Herzenberg LA (2002) Clin Chem 48:1819
28. Givan AL (2004) Methods Mol Biol 263:1
29. Safarik I, Safarikova M (1999) J Chromatogr B 722:33
30. Bonanno G, Perillo A, Rutella S, de Ritis DG, Mariotti A, Marone M, Meoni F, Scambia G, Leone G, Mancuso S, Pierelli L (2004) Transfusion 44:1087
31. Boyum A (1968) Scand J Clin Lab Invest 21(Suppl.97):77
32. Krause DS, Fackler MJ, Civin CI, May WS (1996) Blood 87:1
33. Gordon PR, Leimig T, Babarin-Dorner A, Houston J, Holladay M, Mueller I, Geiger T, Handgretinger R (2003) Bone Marrow Transpl 31:17
34. Ebener U, Brinkmann A, Zotova V, Niegemann E, Wehner S (2000) Klin Padiatr 212:90
35. Yin AH, Miraglia S, Zanjani ED, Almeida-Porada G, Ogawa M, Leary AG, Olweus J, Kearney J, Buck DW (1997) Blood 90:5002
36. Koehl U, Zimmermann S, Esser R, Sørensen J, Gruttner HP, Duchscherer M, Seifried E, Klingebiel T, Schwabe D (2002) Bone Marrow Transpl 29:927
37. Lang P, Bader P, Schumm M, Feuchtinger T, Einsele H, Fuhrer M, Weinstock C, Handgretinger R, Kuci S, Martin D, Niethammer D, Greil J (2004) Br J Haematol 124:72
38. Reth M (1992) Annu Rev Immunol 10:97
39. Delves PM, Roitt IM (2000) N Engl J Med 343:37
40. Berenson RJ, Bensinger WI, Kalamasz D, Martin P (1986) Blood 67:509
41. Berenson RJ, Bensinger WI, Kalamasz D (1986) J Immunol Methods 91:11
42. Kumar A, Plieva FM, Galaev IY, Mattiasson B (2003) J Immunol Methods 283:185
43. Mahe B, Milpied N, Hermouet S, Robillard N, Moreau P, Letortorec S, Rapp MJ, Bataille R, Harousseau JL (1996) Br J Haematol 92:263
44. Lemoli RM, Fortuna A, Motta MR, Rizzi S, Giudice V, Nannetti A, Martinelli G, Cavo M, Amabile M, Mangianti S, Fogli M, Conte R, Tura S (1996) Blood 87:1625
45. Shpall EJ, Jones RB, Bearman SI, Franklin WA, Archer PG, Curiel T, Bitter M, Claman HN, Stemmer SM, Purdy M (1994) J Clin Oncol 12:28
46. Bell GI (1978) Science 200:618

47. Haas W, Schrader JW, Szenberg A (1974) Eur J Immunol 4:565
48. Mammen M, Choi S-K, Whitesides GM (1998) Angew Chem Int Ed 37:2754
49. Dainiak MB, Kumar A, Galaev IY, Mattiasson B (2006) Proc Natl Acad Sci USA 103:849
50. Galaev IY, Dainiak MB, Plieva FM, Mattiasson B (2007) Langmuir 23:35
51. Kumar A, Rodriguez-Caballero A, Plieva FM, Galaev IY, Nandakumar KS, Kamihira M, Holmdahl R, Orfao A, Mattiasson B (2005) J Mol Recog 18:84
52. Hentze H-P, Antonietti M (2002) Rev Mol Biotechnol 90:27
53. Gustavsson P-E, Larsson P-O (1999) J Chromatogr A 832:29
54. Delude RL (2005) Crit Care Med 33:S426
55. Ibrahim SF, Engh Van den G (2003) Curr Opin Biotechnol 14:5
56. McClatchey KD (1994) Clinical laboratory medicine. Williams and Wilkins, Baltimore
57. Sasaki DT, Tichenor EH, Lopez F, Combs J, Uchida N, Smith CR, Stokdijk W, Vardanega M, Buckle AM, Chen B (1995) J Hematother 4:503
58. Ashcroft RG, Lopez PA (2000) J Immunol Method 243:13
59. Chalmers JJ, Zborowski M, Sun L, Moore L (1998) Biotechnol Prog 14:141
60. Beaujean F (1997) Transfus Sci 18:251
61. Kekarainen T, Mannelin S, Laine J, Jaatinen T (2006) BMC Cell Biol 7:30
62. Martin-Henao GA, Picon M, Amill B, Querol S, Ferra C, Granena A, Garcia J (2001) Bone Marrow Transpl 27:683
63. Negrin RS, Atkinson K, Leemhuis T, Hanania E, Juttner C, Tierney K, Hu WW, Johnston LJ, Shizuru JA, Stockerl-Goldstein KE, Blume KG, Weissman IL, Bower S, Baynes R, Dansey R, Karanes C, Peters W, Klein J (2000) Biol Blood Marrow Transpl 6:262
64. Seidl J, Knuechel R, Kunz-Schughart LA (1999) Cytometry 36:102
65. Hu X, Bessette PH, Qian J, Meinhart CD, Daugherty PS, Soh HT (2005) Proc Natl Acad Sci USA 102:15757
66. Kato S, Yabe H, Yasui M, Kawa K, Yoshida T, Watanabe A, Osugi Y, Horibe K, Kodera Y (2000) Bone Marrow Transpl 26:1281
67. Paulus U, Dreger P, Viehmann K, Von Neuhoff N, Schmitz N (1997) Stem Cells 15:297
68. Law P, Ishizawa L, Van de Ven C, Burgess J, Hardwick A, Plunkett M, Gee AP, Cairo M (1993) J Hematother 2:247
69. Melnik K, Nakamura M, Comella K, Lasky LC, Zborowski M, Chalmers JJ (2001) Biotechnol Prog 17:907
70. Belvedere O, Feruglio C, Malangone W, Bonora ML, Donini A, Dorotea L, Tonutti E, Rinaldi C, Pittino M, Baccarani M, Frate GD, Biffoni F, Sala P, Hilbert DM, Degrassi A (1999) Blood Cells Mol Dis 25:140
71. Schumm M, Lang P, Taylor G, Kuci S, Klingebiel T, Buhring H-J, Geiselhart A, Niethammer D, Handgretinger R (1999) J Hematother 8:209
72. Stamm C, Westphal B, Kleine H-D, Petzsch M, Kittner C, Klinge H, Schumichen C, Nienaber CA, Freund M, Steinhoff G (2003) Lancet 361:45
73. Gazitt Y, Reading CC, Hoffman R, Wickrema A, Vesole DH, Jagannath S, Condino J, Lee B, Barlogie B, Tricot G (1995) Blood 86:381
74. Link H, Arseniev L, Bähre O, Kadar JG, Diedrich H, Poliwoda H (1996) Blood 87:4903
75. Bensinger WI, Buckner CD, Shannon-Dorcy K, Rowley S, Appelbaum FR, Benyunes M, Clift R, Martin P, Demirer T, Storb R, Lee M, Schiller G (1996) Blood 88:4132
76. Berenson RJ, Bensinger WI, Hill RS, Andrews RG, Garcia-Lopez J, Kalamasz DF, Still BJ, Spitzer G, Buckner CD, Bernstein ID, Thomas ED (1991) Blood 77:1717
77. Cardoso AA, Watt SM, Batard P, Li ML, Hatzfeld A, Genevier H, Hatzfeld J (1995) Exp Hematol 23:407

Adv Biochem Engin/Biotechnol (2007) 106: 19–39
DOI 10.1007/10_2007_073

Published online: 30 August 2007

Flow Cytometry and Cell Sorting

Sherrif F. Ibrahim[1] (✉) · Ger van den Engh[2,3]

[1]Department of Dermatology, University of Rochester,
601 Elmwood Avenue, Box 697, Rochester, NY 14642, USA
sherrif_ibrahim@urmc.rochester.edu

[2]Department of Oceanography, University of Washington, Seattle, WA USA

[3]Cytopeia, Inc., 12730 28th Avenue NE, Seattle, WA 98125, USA

Abstract Flow cytometry and cell sorting are well-established technologies in clinical diagnostics and biomedical research. Heterogeneous mixtures of cells are placed in suspension and passed single file across one or more laser interrogation points. Light signals emitted from the particles are collected and correlated to entities such as cell morphology, surface and intracellular protein expression, gene expression, and cellular physiology. Based on user-defined parameters, individual cells can then be diverted from the fluid stream and collected into viable, homogeneous fractions at exceptionally high speeds and a purity that approaches 100%. As such, the cell sorter becomes the launching point for numerous downstream studies. Flow cytometry is a cornerstone in clinical diagnostics, and cheaper, more versatile machines are finding their way into widespread and varied uses. In addition, advances in computing and optics have led to a new generation of flow cytometers capable of processing cells at orders of magnitudes faster than their predecessors, and with staggering degrees of complexity, making the cytometer a powerful discovery tool in biotechnology. This chapter will begin with a discussion of basic principles of flow cytometry and cell sorting, including a technical description of factors that contribute to the performance of these instruments. The remaining sections will then be divided into clinical- and research-based applications of flow cytometry and cell sorting, highlighting salient studies that illustrate the versatility of this indispensable technology.

Keywords Flow cytometry · Cell sorting · High-speed cell sorting · Molecular sorting · Clinical diagnostics

1
Introduction

Flow cytometry and cell sorting are vital tools in biological research and clinical diagnostics. Instruments are capable of rapid and highly quantitative interrogation of individual cells for the presence or absence of a wide range of fluorescence and light scattering signals that correlate to cell morphology, surface and intracellular protein expression, gene expression, and cellular physiology [1,2]. From a separations perspective, cell sorting is an indispensable technology, as heterogeneous cell suspensions can be purified into fractions containing a single cell type based upon virtually unlimited combinations of user-defined parameters. Beyond the exceptionally accurate and analytical nature, sorting function is non-destructive to the particles being processed, with little to no effect on cell viability or function. As such, the cell sorter becomes a starting point for boundless cellular and molecular investigations. Moreover, with advances in optics and computing technology, and increasing demand for more intricate, high-throughput analyses, these machines have become powerful discovery tools in biotechnology – facilitating innovation in areas such as protein engineering, drug screening, and cell signal profiling [3]. The newest and relatively nascent arena for cell sorting relates to medical therapy. While clinical diagnostics is inextricably tied to flow cytometry, cell based therapeutics have not reached mainstream practice. Assuredly, this will be an area of expansion in the future.

Although flow cytometry and cell sorting are established, mature technologies [4,5], the evolution of new generation instrumentation continues along a dichotomous course. Demands of the new biology require machines to function at higher speeds, capable of increasing experimental complexity, while the drive for better and more practical clinical diagnostic tests imposes demands for cheap and robust analysis, particularly in geographic areas where expensive machines are not feasible, and monitoring of disease and efficacy of therapies is critical [6–8].

This chapter will begin with an overview of flow cytometry and cell sorting, providing the reader with a basic background in cytometer function. To follow we will discuss how changes in optics, fluidics, and computing are influencing the field's progression, focusing on specific areas that this technology has had a great impact upon.

2
Principles of Flow Cytometry

In essence, a flow cytometer is a fluorescence microscope that collects light signals emitted from particles flowing across an objective lens [1,2]. There are several styles of fluidics configurations, and this chapter will focus on the jet-in-air type, as this represents the vast majority of available systems. In all cytometers, the cells or particles to be analyzed require being placed in suspension and pressurized into a directed fluid stream. To ensure that particles emerge in single file for analysis, the sample is injected as a core stream into a larger, surrounding column of flowing sheath fluid. Under conditions of laminar flow, the sample then becomes hydronamically focused within the sheath stream. Air or another gas pressurizes both streams, and it is the differential pressure between the streams that controls sample injection rate. Particles then emerge in single file from a small nozzle (typically 70 μm in diameter) to form a narrow jet of fluid in air. This stream containing particles sequentially intersects one or more laser beams placed orthogonal to the flow of fluid. The laser beams are focused such that they only illuminate a single particle at any given time. If the given cell or particle contains a fluorescent tag that is excitable by the laser, a measurable pulse of photons of a specific wavelength will be emitted. Signals are collected by an array of photo-detectors and optical filters, processed by specialized electronics, and stored on a computer. At constant laser power, the intensity of emission will be dependent on the number of fluorophores present, thereby making flow cytometry both a qualitative and highly quantitative analysis tool. By using multiple lasers of different wavelengths, the potential number of excitable fluorophores is increased, as fluorescence signals are emitted sequentially from the same particle intersecting each laser.

Figure 1 depicts the process of flow cytometry. Sheath fluid with a central core containing the sample to be analyzed is seen emerging from the nozzle at the top of the figure. As the stream intersects the two lasers shown, light signals are collected by the objective lens. Downstream from the analysis point, the stream is broken into discrete droplets so that desired particles can be diverted from the stream while the remainder are disposed of. The details of cell sorting will be described in further detail below. While the most basic machines use a single illumination laser, more advanced instruments can use up to six excitation paths (Fig. 2). Because the jet is flowing at a constant velocity, signals are separated by a constant time. With appropriate cytometer calibration, these signals are then processed and displayed to correspond to an individual particle. In addition to fluorescence signals, incident laser light is also scattered from the flowing cells. While fluorescence emission will always be at a longer wavelength (lower energy) than the laser, scattered light will be at the same wavelength. Light scattered in the forward direction (at a small angle with respect to the incident laser light) is proportional to the size of the particle, while light scattered perpendicularly (orthogonal to the incident laser light) correlates with intracellular granularity. Thus, as a result of light scattering data alone, one can elicit useful information regarding cell morphology.

Both fluorescence and scatter data are typically displayed in the form of univariate histograms or bivariate scatter plots by specialized software

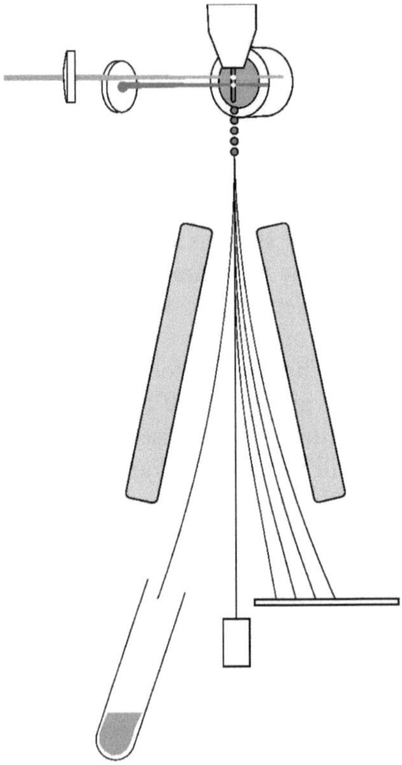

Fig. 1 Principles of flow cytometry. Particles are placed in a fluid stream and pass by a laser in single file. Light scattering and fluorescence signals are detected for each passing particle. Sheath fluid with a central core containing the sample to be analyzed is seen emerging from the nozzle at the top of the figure. As the stream intersects the two lasers shown, light signals are collected by the round objective lens. Downstream from the analysis point, the stream is broken into discrete droplets so that desired particles can be charged and diverted from the stream in an electric field while the remainder are disposed of

(Fig. 3). In the case of histograms, the x-axis will represent either fluorescence or scatter intensity collected from a single photo-detector on either a linear or logarithmic scale, and the y-axis represents the number of particles with the corresponding light intensity. For instance, if a cell is tagged with a fluorescently labeled antibody directed towards a surface protein, the fluorescence intensity will be directly proportional to the expression level of this protein. By using multiple antibodies one could then assess the expression levels of several membrane-bound and/or intracellular proteins of a single cell. Any two of these parameters can be displayed simultaneously on bivariate plots displaying data from hundreds to thousands of cells, providing a visual display that highlights different populations of cells. Because data are easily collected from hundreds of thousands to millions of individ-

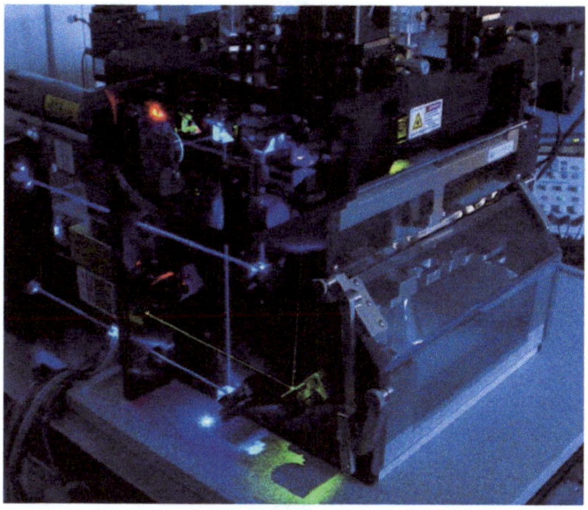

Fig. 2 High-performance cell sorter demonstrating six excitation paths. Individual laser beams are seen being steered through the illumination light path before being focused on the particle stream. Additional laser beams allow the measurement of more parameters for each individual cell

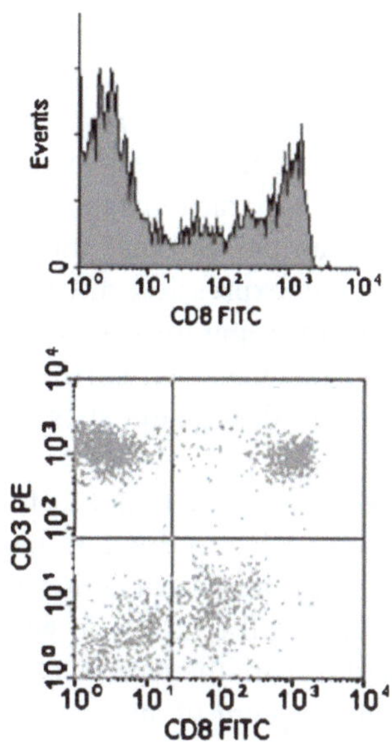

Fig. 3 Sample display of data from a flow cytometer. The *lower figure* is a two-dimensional scatter plot with each dot corresponding to a single cell and antigens expressed on its surface. The *upper figure* represents the x-axis from the scatter plot displayed in histogram format

ual cells, flow cytometry provides tremendous statistical power in a short amount of time. Through the input of user-defined thresholds (commonly referred to as "gates"), one can determine the percentage of cells with a cer-

Fig. 4 Image of the drops generated by introducing an acoustic wave to the fluid stream. The break-off point is extremely stable allowing for the precise charging of individual drops and deflection of desired particles

tain characteristic or combination of features (e.g., protein or gene expression). Depending on the specific instrument being used, the operator then has the remarkable ability to separate ultrapure populations of cells from the original, heterogeneous sample. Because the end point of many studies is simply to obtain data on the expression of certain genes and/or proteins without the need to physically separate the cells, there is a large market for flow cytometers that do not have cell sorting capability (referred to as analyzers). Machines used in clinical diagnostics typically belong to this category [1, 2].

As the name implies, cell sorters, as opposed to pure analyzers, have the added benefit of being able to select individual particles of interest and divert them from the fluid stream into a collection vessel. This is achieved by introducing a slight vibration on the nozzle to generate small waves on the surface of the jet as it emerges from the nozzle, causing it to break into regular droplets downstream of the point of laser intersection (Fig. 4). If a given particle is desired by the operator (as defined by a certain combination of fluorescent and light scattering parameters), an electric charge is applied to those drops containing the particles of interest. These particle-containing drops can then be deflected in an electric field and collected while the remaining, uncharged drops are disposed of. The amount of charge applied will affect the degree of deflection of individual drops, and hence, multiple populations of cells can be separated simultaneously by using charges of different polarity and intensity. Because the electric charges on any given drop will repel each other, the drop of sheath fluid enclosing a desired particle acts as a Faraday cage, protecting the contained cell from the effects of the electric charge.

3 High-Speed Cell Sorting

With a basic understanding of flow cytometry and cell sorting, one can begin to ask questions about the factors governing the performance of these machines. In particular, how fast can they operate without compromising the integrity of collected data or purity of sorted populations? As modern scientific research is focused on system-based approaches, high-throughput analysis becomes an absolute requirement, and cell sorting technology has evolved accordingly.

Because each particle is analyzed individually, the performance of cell sorters is intrinsically slow. The limiting rate by which particles can be analyzed and then sorted is dictated by basic physical and stochastic principles. Concurrent advances in fluidics, optics, computers, and the electronics of sorters have led to the development of instruments referred to as high-speed cell sorters that can analyze and sort cells at rates approaching two orders of magnitude over previous models. When designing a flow cytometer capable of operating at these higher rates, it is necessary to optimize each component of the sorting process without compromising sort quality and yield. For example, manipulation of fluidics will have direct effect on how signals are received and subsequently processed, and vice versa.

As particles are illuminated by one or more laser beams, specialized circuits receive signals sent from the photo-detectors, perform analog-to-digital conversion, classify the events, and issue sort commands for those cells that fall within specified sort windows. Modern high-level data acquisition systems and analog-to-digital converters are capable of processing signals on the order of 1 MHz, and are therefore not limiting to the overall sort process. The electronic circuitry of modern high-speed cytometers is digital at heart with the best machines having little problem quantifying and classifying upwards of 16 measurement parameters in the time it takes a cell to cross the laser interrogation points [9]. However, it is not possible to sort events at a higher rate than that at which sort drops are generated. Therefore, to achieve the highest possible sort rate, the speed of drop generation should approximate the speed of the electronics.

The speed of drop generation is subject to the laws of fluid dynamics. The maximum drop formation rate is proportional to the jet velocity, which in turn is proportional to the square root of the jet pressure [10]. This relationship predicts that the generation of 250×10^3 drops s^{-1} by a 70 μm nozzle would require a jet pressure of approximately 500 psi [2], higher than what cells can sustain to remain viable during their transit from jet to collection vessel, not to mention outside the realm of safe laboratory operating procedure. This pressure limitation translates to a reasonable maximum drop generation rate of approximately $100 \times 10^3\ s^{-1}$, or a jet velocity of roughly $25\ m\,s^{-1}$, and it is indeed in this regime where the top high-speed flow cytometers perform. Because of this exponential pressure/velocity relationship, the increased pressure at which high-speed flow cytometers operate does not easily translate into increased fluid velocity, or simply "pushing" events across the laser beams at a higher rate. While higher jet velocity does indeed result in decreased transit time across the laser beams (thereby decreasing the signal pulse duration and freeing up the electronics to receive the next signal), substantial increases in velocity are only achieved with tremendous and, to some extent, unattainable increases in pressure [11].

Laws of probability also play a role in determining maximum sort rates. Given a stable drop formation system and robust electronics, if the particles to be analyzed arrive at the interrogation point at precisely the same interval as the drop formation rate, then the maximum sort rate approximates the rate of drop formation. However, as stochastic principles dictate, a high event rate will ultimately lead to losses in effective sort rate due to coincident drop occupancy. In other words, when two cells are in the beam simultaneously the signal cannot be interpreted unequivocally. Likewise, there will be a certain frequency of drops containing no particles. In these situations, when the sorter is uncertain about the accuracy of the measurement, it must abort sort operations. At both an event rate and drop formation rate of 100 KHz, taking into account abort rates based on coincident drop occupancy, the maximum sort rate is more along the order of 40×10^3 particles s^{-1} [1]. The discussion above reinforces the argument that the term *high-speed* does not refer to simply speeding up of the sort process, but is the culmination of precise engineering of each individual component in order to optimize overall performance. In addition to these advancements, modern high-end cytometers have also become modular, with openly accessible and arrangeable parts, allowing for greater flexibility in experimental design and more stable operation.

4 Clinical Diagnostics

For analysis purposes, the faster sort rates achievable by high-speed sorters are not as important, as only a small number of cells need to be analyzed to determine population distributions. However, the increased capabilities and flexibility of high-speed sorters to measure many parameters in parallel using multiple lasers with high sensitivity and accuracy has greatly benefited both researchers and clinicians alike [12–17]. This ability to extract more information per passing particle coupled with the ability to isolate desired cells at relatively high throughput make high-speed cell sorters a powerful and versatile technology. Whereas the earliest flow cytometric analyses looked for the presence of a single marker against a negative background, it is common today to classify certain hematologic disorders with as many as ten different fluorescence parameters [12, 18]. This number will undoubtedly be surpassed with improvements in cell sorting technology and biological discovery methods. Many of these types of analyses that rely on fine discrimination between various cell types were not possible until the advent of high-speed flow cytometry. And, as the ability to classify cells increases, so too will the arsenal of methods available to label cells.

While high-speed cytometers are becoming increasingly sophisticated for the reasons mentioned above, the evolution of flow cytometry is also progressing towards cheaper, simpler machines that can be used for routine analysis and can be operated by users with little technical training, with the vast majority of flow cytometers currently in use geared towards clinical diagnostics. With increasing financial constraints from healthcare insurance providers, these machines are being manufactured at lower prices and geared for use in settings outside core facilities. For example, developments in the use of light-emitting diodes (LED) appears to be a promising avenue for price reduction in less expensive instruments [19]. Perhaps the greatest area where the impact of cheaper cytometers has been felt is in the detection and monitoring of human immunodeficiency virus (HIV) infection and its progression to the acquired immunodeficiency syndrome (AIDS). Disease status is monitored by the patients CD4 T-helper cell counts, a parameter simply measured in a cytometer. With flow cytometry, AIDS monitoring was first possible in 1981 [20]. In a 2004 report on the status of HIV diagnostics, it was noted that in Brazil alone there were 85 flow cytometers in the field, with an estimated cost of CD4 count per sample of less than five dollars [8, 21].

Blood provides an ideal substrate for cytometry, as cells are provided in suspension. Standard in any hospital laboratory are machines based on principles of flow cytometry that can enumerate the complete blood count (CBC), allowing rapid insight into factors such as patient's red blood cells, white blood cells, monocytes, eosinophils, and platelets – each with its own corollaries into a patient's health status. Multiparametric flow analysis is quickly replacing fluorescence microscopy and immunocytochemistry techniques in the diagnosis and classification of many hematologic disorders and evaluation of graft quality [18, 22–26]. Cells of different lineages differ morphologically, and hence are easily distinguishable by light scattering properties alone. Discrete populations of cells can be purified from a heterogeneous mixture for determination of their surface proteins. This type of analysis has certainly transformed the field of hematology, initially in the establishment of the CD nomenclature, and today with standard use of flow in the study of myelodysplastic syndromes [18]. Flow cytometry is used routinely for diagnosis, staging, classification, degree of involvement (i.e., cerebral spinal fluid analysis), and detection of minimal residual disease. Within the last few years alone, reports have appeared in the literature describing cellular analysis across a broad range of clinical fields including immunology, oncology, hematology, blood banking, transplantation medicine, and genetics [27–34].

Prenatal diagnostics is another clinical area where flow cytometry will continue to have an impact. Current methods rely on amniocentesis or chorionic villus sampling, procedures which are not without risk. More recently, there have been reports of using flow cytometry to identify and isolate fetal cells within maternal blood [35]. This provides an easy, relatively safe, and inexpensive approach for prenatal diagnostics. Assessment of sperm quality by flow cytometry has also been reported [36].

Additionally, as flow cytometers become cheaper, more manageable, and easier to operate, they may become portable for field use in a wide assortment of immunological and infectious screening applications. There will undoubtedly be increased use of cytometers in field clinics, water monitoring, agriculture/veterinary diagnostics, and rapidly deployable biothreat detection [37]. A recent study was able to use flow cytometry to detect changes related to chronic respiratory inflammation resulting in workers exposed to inhalational toxins [38]. It is conceivable that smaller, handheld units may even be employed for use in the battlefield during wartime [39]. More recent, proven applications for flow cytometry demonstrating their increasing utility and portability relate to microbial and oceanographic applications, with flow cytometers now present on several oceanographic research vessels and revealing tremendous information regarding biodiversity in relatively unstudied systems such as the soil and the sea [40–42].

5 Cell-Based Therapy

To date, clinical flow cytometric studies have been mainly limited to diagnostic, monitoring, and classification purposes. As the reliability of cell sorting improves, there has been an emergence of cell sorters as a means for cell-based therapy [1]. In this regard, high-speed flow cytometers may make the most impact, as these applications demand the greatest need for higher throughput and flexibility in measurement parameters; cells to be infused into a patient should be extracorporeal for the shortest time possible and isolated to the utmost level of purity. For instance, in the case of bone marrow transplantation, it was unknown which of the transplanted cells had the major regenerative capacity. As a result, patients would receive transplants of whole HLA-identical sibling bone marrow and hope for the best outcome. HLA-mismatched, or unrelated grafts with concurrent immunosuppression raised issues of tolerance and graft vs. host disease (GVHD). In 1984, Martin et al. published a study reporting the use of "microfluorometry" to deplete T-cells from donor marrow in order to decrease GVHD [43]. In the 1990s, the discovery of the $CD34^+$ phenotype led clinicians to believe that cells displaying this surface antigen contained the stem cell function of bone marrow, and it was soon shown that as few as 7×10^6 $CD34^+$cells per kilogram would be sufficient for engraftment [44]. Cells could be labeled with a CD34-reactive antibody and purified in a cell sorter. Additional advances in hematology and immunophenotyping will undoubtedly lead to more successful transplantations with fewer complications [45, 46].

Within the last few years, the rapidly expanding set of genes known to play a role in cellular differentiation has led to complex staining patterns [47–49], or "molecular signatures" [50] characteristic of the most primitive cells. So while the first applications of clinical flow cytometry relied on screening through large numbers of cells to acquire sufficient material for successful engraftment, more recent trials depend on the successful identification of a relatively rare population of uniquely labeled cells in the peripheral blood, marrow, or cord blood compartment. While fewer of these highly defined cells are needed, speed is of the essence, as by definition a more highly classified cell will be present in lesser quantities. As we better determine which cells are crucial to this and other processes, high-speed cell sorters will be at the forefront of many treatment regimes. A natural extension of the use of high-speed cell sorters in the setting of cancer treatment is for tumor cell purging. As our knowledge of the landscape of healthy cells increases, so too does the information regarding malignant cells [51, 52]. For instance, if the phenotype of a known B- or T-cell leukemia is known, cells can be removed from the patient, treated ex vivo to remove cells bearing this combination of markers and reinfused into the patient [53].

As an additional consideration, high-speed cell sorting for clinical applications raises aspects of sorting not encountered with experimental protocols in the laboratory. Issues of absolute sterility, cleanliness, and reliability are of utmost importance, both to prevent sample contamination and to protect the sort operator from aerosolized particles. Further examples of cell sorters for cell-based therapy beyond hematologic disorders include gene therapy [52], pancreatic islet cell transplantation [54], and sperm sorting for gender preselection [55]. The latter is the area where sorting of mammalian cells for reintroduction to a living organism is most used and is established with the Food and Drug Administration (FDA). Originally devised for the selection of gender-specific livestock [55], the technique relies on the differential DNA contact of X and Y chromosome-bearing sperm. Because the X chromosome is significantly larger than the Y chromosome, when

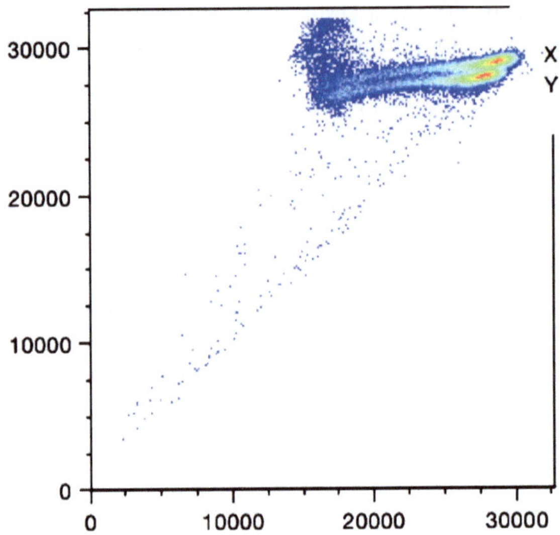

Fig. 5 Scatter plot of bull sperm stained with a generic DNA dye. The X-bearing sperm contain more DNA, therefore appearing as a separate population from Y-bearing sperm

sperm are incubated with a fluorescent dye that generically binds to DNA, the fluorescence intensity is considerably higher from X chromosome-bearing sperm than that from those sperm containing a Y chromosome (Fig. 5). As such, X-bearing sperm can be separated in bulk and used for in vitro fertilization to greatly enhance the chances for female offspring. In humans, the difference in DNA content between X- and Y-bearing sperm accounts for an approximately 2% difference in overall DNA content, and hence fluorescence signal detected by flow cytometry. Indeed this method has been employed for the use of gender-selection in the case of families with known X-linked diseases, and to assist with "family balancing", sparking ongoing ethical debate.

6 High-Speed Cell Sorting in the Research Laboratory

Cell sorting allows the investigator to analyze quantitatively several fluorescence and light scattering parameters of individual particles and to purify those events with the desired characteristics for further study. No other technology can separate a heterogeneous cell suspension into purified fractions containing a single cell type with the speed and accuracy of high-speed cell sorters. While the original development of high-speed sorters was driven by the need for purified chromosomes in the early stages of the human genome project [56, 57], their current uses cover a vast spectrum of biological research. Applications for high-speed cell sorting are commonly grouped into two categories: bulk sorting and rare event sorting.

In bulk sorting, the sort operator is given a large number of cells of which a small percentage is expressing a given measurable phenotype and this population or populations are then used for further study. As examples, this marker can be a surface antigen bound to a fluorescently labeled antibody, a transient transfection expressing a given protein of interest, activated leukocytes, or any detectable measure of cell state or type. Whatever the measured parameter, the operator must screen through a large number of cells in order to collect a sufficient amount of material for the next step of analysis. For instance, an investigator may be interested in the top 3% of cells expressing a given intracellular cytokine in order to perform proteomic analysis. Typically, a fairly large number of cells are needed for these studies. With readily available mass spectrometry methods, one would need on the order of 10^7 cells to detect a protein expressed at roughly 1000 copies per cell, corresponding to a total starting population of approximately 2×10^8 cells. At a sort rate of 40 000 cells s^{-1} this translates to 5000 s or about an hour and a half of sorting. This does not take into account material losses due to sample handling (e.g., centrifugation, etc.) or time losses due to technical issues. Realistically, perhaps twice as many cells or twice as much overall time would be needed to achieve the desired number. However, even doubling the sort time would not be a major issue, easily falling within the realm of a typical day in the laboratory. When comparing these figures with a standard flow cytometer capable of sort rates of approximately 1000 cells s^{-1}, the benefits of high-speed sorting immediately become evident, if not critical to the process [1]. Users of traditional cell sorters can overcome this problem to some extent via a pre-enrichment step such as magnetic bead separation prior to sorting. However, this is currently only possible when selecting cells based on surface markers, not for intracellular stains or products of fluorescent reporter genes, or for combinations of multiple parameters.

The highly quantitative, multiparametric nature of flow cytometry and cell sorting has played an integral role in the understanding of many biological processes. Each cell in a given population of cells will respond differently to a given stimulus resulting in a full spectrum of responses. Cells in a given study will map to different regions on the cell sorter computer screen as it displays selected parameter values. Methods of systematically isolating and comparing these subpopulations of cells are responsible for many discoveries of both normal and pathologic cellular processes. Studies of purified cells by themselves or in combination can reveal information regarding cell function. As an example, in a report by Liu et al., it was shown that cells separated on the basis of differentially expressed surface markers acted differently when isolated or in unison with other cell types in directing and maintaining prostatic cellular differentiation [58]. Much of the early work outlining hematopoiesis was achieved in this fashion and eventually contributed to the first isolations of the hematopoietic stem cell [59]. Currently, cell sorters continue to be at the crux of stem cell research [60, 61] and cell type-specific developmental analysis [62]. One would be hard-pressed to find a many publications relating to stem cells that did not employ flow cytometry and cell sorting as part of their methods [49, 61]. In a recent report, epidermal skin stem cells were isolated in a flow cytometer based on surface markers, and their nuclei were reprogrammed to grow mice by their exposure to the cytoplasm of unfertilized murine oocytes [63]. Other studies have used the flow cytometer as a tool for dynamic measurement of dye uptake kinetics to help identify progenitor cells [64, 65].

7 Molecular Sorting

A second rapidly expanding application of high-speed flow cytometry is rare event sorting. For clinical applications, this translates into screening through large numbers of cells to obtain a more defined, lower frequency cell necessary to facilitate a given course of treatment. Recently, however, rare event sorting has referred to reading through highly complex cell-based or solid-support molecular libraries in order to select genetic clones and/or molecules of interest for further chemical or molecular studies. Flowing particles become vehicles encoding bits of information for the cell sorter to decipher according to operator-defined parameters. With protein expression library screening, cells provide a physical link between expressed phenotype and genotype [1]. Ingenious fluorescent labeling schemes are used to indicate binding or activity characteristic of a specific outcome both qualitatively and quantitatively [66, 67]. These cells can then be repeatedly sorted and expanded, thereby enriching for desired clones.

The cloning of fluorescent reporter genes such as green fluorescent protein, dsRed, and their homologs has revolutionized the field of fluorescence microscopy and flow cytometry through their ability to tag individual cells for the expression of single genes or a combination of genes [68–71]. The expression of mammalian and bacterial proteins can be directly monitored by fluorescent reporters easily detectable with mainstream flow cytometers. The discussion of fluorescent proteins in and of itself is the subject of entire textbooks and relies on the ability to genetically link DNA encoding for the fluorescent protein with, theoretically, any other protein in the system. Monitoring of gene expression and protein localization were but the first uses for this technology. Outlined below are some of the more ingenious studies using these and other fluorescent labeling schemes.

Techniques such as periplasmic expression with cytometric screening (PECS) [72] and yeast surface display [73] are excellent examples of methods that combine expression library technology and cell sorting. Using the latter, Boder et al., were able to isolate a single chain antibody with femtomolar antigen-binding affinity, the highest ligand-binding affinity for an engineered protein [74]. A study by Olsen et al. used bacterial surface display and a clever fluorescence-based method to cytometrically select protein variants based on enzymatic catalytic turnover [75].

In the case of DNA-based libraries, a fluorescent signal may indicate the occurrence of a particular genetic event of interest. The high-speed cell sorter then becomes a tool to potentially screen through entire genomes of organisms in a relatively short period of time. Clones displaying a certain phenotype can be sorted, their DNA PCR-amplified, and sequenced to identify sequences of note. In the case of *Saccharomyces cerevisiae* with a genome size of approximately 14 Mbp, 100 bp fragments could be generated and placed upstream of a reporter construct, creating a library on the order of 10^5 clones. The entire genome could be scanned to 10X coverage for the occurrence of a particular genetic event in a matter of minutes using high-speed cell sorters. As an example, a study by Barker et al. generated a library of 200–1000 bp fragments of the *Mycobacterium marinum* genome fused to a promoterless copy of green fluorescent protein (GFP). Like its counterpart *Mycobacterium tuberculosis*, this bacterium is able to avert the normal function of the immune system and survive within macrophages. By isolating phagosomes emitting green fluorescence and sequencing the regions of DNA upstream of GFP, they were able to determine which regions of the genome may be responsible for circumventing normal bacterial killing [76]. Again, fluorescent reporter genes will play an ever-increasing role in these types of rare particle, or perhaps more aptly termed, rare "occurrence" studies.

Other recent reports of note include the use of flow cytometers to screen for G-protein-coupled receptors [77], to decipher immunologic signaling networks with phosphospecific flow cytometry [78] and other enzymatic activity [79], high-throughput drug screening and signaling through the use of bar-coding [80], and the cell sorter as a DNA sequencing and transcriptional profiling tool [81]. A recent study used a dual fluorescent protein reporter to select for plasmids containing DNA inserts as templates for sequencing as a method for bacterial clone selection without plating. By designing the insertion site between the genes encoding for two different fluorescent proteins, insertion of genetic material resulted in loss of expression of the downstream fluorescent protein (Fig. 6) [82]. Several recent reviews outline other ingenious applications for flow cytometry and cell sorting in the arena of biotechnology as well as discuss changes in the technology behind these instruments [1, 3, 83–91].

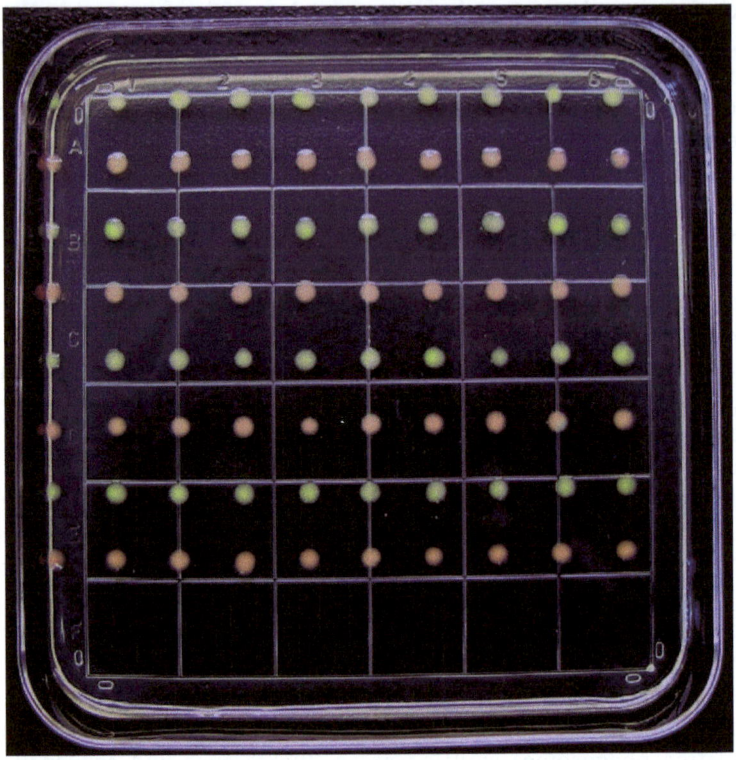

Fig. 6 Individual *E. Coli* sorted directly on to an agar plate based on expression of GFP vs. dsRED and allowed to incubate for 36 h. Single bacterial cells were sorted directly on to the agar plate, alternating rows of cells expressing GFP and dsRed. After incubation, discrete pure colonies of cells expressing the respective fluorescent protein are seen. This displays the extreme accuracy of sort commands of a single cell and the power of fluorescent protein technology. Photograph courtesy of Dr. J. Choe

8 Conclusions

Flow cytometric-based cell sorting is a well-established technology that will continue to see expansion of its uses in many areas, including clinical diagnostics, medical therapy, and scientific research. Cheaper, more portable machines continue to emerge that allow the use of these instruments in essentially any setting. Continued advances in lasers, fluidics, optics, and computing, along with greater molecular and cellular labeling techniques will allow for more accurate, higher throughput machines capable of finer discrimination between cell types [92, 93]. Ultimately, this will lead to a more precise understanding of biology, and contribute to our increasingly detailed view of processes at the single cell level.

References

1. Ibrahim SF, van den Engh G (2003) High-speed cell sorting: fundamentals and recent advances. Curr Opin Biotechnol 14(1):5–12
2. van den Engh G (2000) High-speed cell sorting. In: Durack G, Robinson JP (eds) Emerging tools for single-cell analysis: advances in optical measurement technologies. Wiley-Liss, New York
3. Mattanovich D, Borth N (2006) Applications of cell sorting in biotechnology. Microb Cell Fact 21(5):12
4. Herzenberg LA, Parks D, Sahaf B, Perez O, Roederer M, Herzenberg LA (2002) The history and future of the fluorescence activated cell sorter and flow cytometry: a view from Stanford. Clin Chem 48:1819–1827
5. Herzenberg LA, Sweet RG, Herzenberg LA (1976) Fluorescence-activated cell sorting. Sci Am 234:108–117
6. Baumgarth N, Roederer M (2000) A practical approach to multicolor flow cytometry for immunophenotyping. J Immunol Methods 243:77–97
7. Ashcroft RG, Lopez PA (2000) Commercial high speed machines open new opportunities in high throughput flow cytometry (HTFC). J Immunol Methods 243:13–24
8. Mandy FF (2004) Twenty-five years of clinical flow cytometry: AIDS accelerated global instrument distribution. Cytometry A 58(1):55–56
9. van den Engh G, Stokdijk W (1989) Parallel processing data acquisition system for multilaser flow cytometry and cell sorting. Cytometry 10:282–293
10. Stovel RT (1977) The influence of particles on jet breakoff. J Histochem Cytochem 25:813–820
11. van den Engh G, Farmer C (1992) Photo-bleaching and photon saturation in flow cytometry. Cytometry 13:669–677
12. Wood B (2006) 9-color and 10-color flow cytometry in the clinical laboratory. Arch Pathol Lab Med 130(5):680–690
13. Pagnucco G, Vanelli L, Gervasi F (2002) Multidimensional flow cytometry immunophenotyping of hematologic malignancy. Ann NY Acad Sci 963:313–321
14. Jennings CD, Foon KA (1997) Recent advances in flow cytometry: application to the diagnosis of hematologic malignancy. Blood 90:2863–2892
15. Alvarez-Barrientos A, Arroyo J, Canton R, Nombela C, Sanchez-Perez M (2000) Applications of flow cytometry to clinical microbiology. Clin Microbiol Rev 13:167–195
16. De Rosa SC, Roederer M (2001) Eleven-color flow cytometry. A powerful tool for elucidation of the complex immune system. Clin Lab Med 21:697–712
17. O'Gorman MR (2001) Clinically relevant functional flow cytometry assays. Clin Lab Med 21:779–794
18. Kaleem Z (2006) Flow cytometric analysis of lymphomas: current status and usefulness. Arch Pathol Lab Med 130(12):1850–1858
19. Telford WG (2004) Small lasers in flow cytometry. Methods Mol Biol 263:399–418
20. Gottlieb MS, Schroff R, Schanker HM, Weisman JD et al. (1981) *Pneumocystis carinii* pneumonia and mucosal candidiasis in previously healthy homosexual men: evidence of a new acquired cellular immunodeficiency. N Engl J Med 305(24):1425–1431
21. Hengel RL, Nicholson JK (2001) An update on the use of flow cytometry in HIV infection and AIDS. Clin Lab Med 21:841–856
22. Gutensohn K, Magens MM, Kuehnl P (2000) Flow cytometry: the standard for monitoring the onset of apheresis and for the evaluation of stem and progenitor cell graft quality. Vox Sang 78(suppl 2):131–135

23. Papadaki HA, Marsh JC, Eliopoulos GD (2002) Bone marrow stem cells and stromal cells in autoimmune cytopenias. Leuk Lymphoma 43:753–760
24. Ward MS (1999) The use of flow cytometry in the diagnosis and monitoring of malignant hematological disorders. Pathology 31:382–392
25. Bene MC, Bernier M, Castoldi G, Faure GC, Knapp W, Ludwig WD, Matutes E, Orfao A, van't Veer M (1999) Impact of immunophenotyping on management of acute leukemias. Haematologica 84:1024–1034
26. Brown M, Wittwer C (2000) Flow cytometry: principles and clinical applications in hematology. Clin Chem 46:1221–1229
27. Miller DT, Stelzer GT (2001) Contributions of flow cytometry to the analysis of the myelodysplastic syndrome. Clin Lab Med 21:811–828
28. Stetler-Stevenson M, Braylan RC (2001) Flow cytometric analysis of lymphomas and lymphoproliferative disorders. Semin Hematol 38:111–123
29. Bleesing JJ, Fleisher TA (2001) Immunophenotyping. Semin Hematol 38:100–110
30. Horsburgh T, Martin S, Robson AJ (2000) The application of flow cytometry to histocompatibility testing. Transpl Immunol 8:3–15
31. Bray RA (2001) Flow cytometry in human leukocyte antigen testing. Semin Hematol 38:194–200
32. Escribano L, Diaz-Agustin B, Bellas C, Navalon R, Nunez R, Sperr WR, Schernthaner GH, Valent P, Orfao A (2001) Utility of flow cytometric analysis of mast cells in the diagnosis and classification of adult mastocytosis. Leuk Res 25:563–570
33. Michels JJ, Duigou F, Marnay J (2000) Flow cytometry in primary breast carcinomas. Prognostic impact of proliferative activity. Breast Cancer Res Treat 62:117–126
34. Levine JE, Braun T, Penza SL, Beatty P, Cornetta K, Martino R, Drobyski WR, Barrett AJ, Porter DL, Giralt S et al. (2002) Prospective trial of chemotherapy and donor leukocyte infusions for relapse of advanced myeloid malignancies after allogeneic stem-cell transplantation. J Clin Oncol 20:405–412
35. Purwosunu Y, Sekizawa A, Koide K, Okazaki S, Farina A et al. (2006) Clinical potential for noninvasive prenatal diagnosis through detection of fetal cells in maternal blood. Taiwan J Obstet Gynecol 45(1):10–20
36. Graham JK (2001) Assessment of sperm quality: a flow cytometric approach. Anim Reprod Sci 68:239–247
37. Petrenko VA, Sorokulova IB (2004) Detection of biological threats. A challenge for directed molecular evolution. J Microbiol Methods 58(2):147–168
38. Lyapina M, Zhelezova G, Petrova E, Boev M (2004) Flow cytometric determination of neutrophil respiratory burst activity in workers exposed to formaldehyde. Int Arch Occup Environ Health 77(5):335–340
39. Shapiro HM, Perlmutter NG (2006) Personal cytometers: slow flow or no flow? Cytometry A 69(7):620–630
40. Paul J, Scholin C, van den Engh G, Perry MJ (2007) In-situ instrumentation. Oceanography (sect III) 20(2):58–66
41. Nielsen JL, Schramm A, Bernhard AE, van den Engh GJ, Stahl DA (2004) Flow cytometry-assisted cloning of specific sequence motifs from complex 16S rRNA gene libraries. Appl Envir Microbiol 70:7550–7554
42. Legendre L, Courties C, Troussellier M (2001) Flow cytometry in oceanography 1989–1999: environmental challenges and research trends. Cytometry 44:164–172
43. Martin PJ, Hansen JA, Thomas ED (1984) Use of flow microfluorometry in bone marrow transplantation. Ann NY Acad Sci 428:14–25

44. Perez-Simon JA, Caballero MD, Corral M, Nieto MJ, Orfao A, Vazquez L, Amigo ML, Berges C, Gonzalez M, Del Canizo C, San Miguel JF (1998) Minimal number of circulating CD34+ cells to ensure successful leukapheresis and engraftment in autologous peripheral blood progenitor cell transplantation. Transfusion 38:385–391
45. Sasaki DT, Tichenor EH, Lopez F, Combs J, Uchida N, Smith CR, Stokdijk W, Vardanega M, Buckle AM, Chen B et al. (1995) Development of a clinically applicable high-speed flow cytometer for the isolation of transplantable human hematopoietic stem cells. J Hematother 4:503–514
46. Lamb LS Jr (2002) Hematopoietic cellular therapy: implications for the flow cytometry laboratory. Hematol Oncol Clin North Am 16:455–476
47. Weissman IL (2002) The road ended up at stem cells. Immunol Rev 185:159–174
48. Goodell MA, Brose K, Paradis G, Conner AS, Mulligan RC (1996) Isolation and functional properties of murine hematopoietic stem cells that are replicating in vivo. J Exp Med 183:1797–1806
49. Wognum AW, Eaves AC, Thomas TE (2003) Identification and isolation of hematopoietic stem cells. Arch Med Res 34(6):461–475
50. Ivanova NB, Dimos JT, Schaniel C, Hackney JA, Moore KA, Lemischka IR (2002) A stem cell molecular signature. Science 298:601–604
51. Eliopoulos N, Al-Khaldi A, Beausejour CM, Momparler RL, Momparler LF, Galipeau J (2002) Human cytidine deaminase as an ex vivo drug-selectable marker in gene modified primary bone marrow stromal cells. Gene Ther 9:452–462
52. Dirven CM, Grill J, Lamfers ML, Van der Valk P, Leonhart AM, Van Beusechem VW, Haisma HJ, Pinedo HM, Curiel DT, Vandertop WP, Gerritsen WR (2002) Gene therapy for meningioma: improved gene delivery with targeted adenoviruses. J Neurosurg 97:441–449
53. Shan D, Ledbetter JA, Press OW (1998) Apoptosis of malignant human B cells by ligation of CD20 with monoclonal antibodies. Blood 91:1644–1652
54. Jindal RM, McShane P, Gray DW, Morris PJ (1994) Isolation and purification of pancreatic islets by fluorescence activated islet sorter. Transplant Proc 26:653
55. Johnson LA, Welch GR, Keyvanfar K, Dorfmann A, Fugger EF, Schulman JD (1993) Gender preselection in humans? Flow cytometric separation of X and Y spermatozoa for the prevention of X-linked diseases. Hum Reprod 8:1733–1739
56. Gray JW, Dean PN, Fuscoe JC, Peters DC, Trask B, van den Engh GJ, Van Dilla MA (1987) High-speed chromosome sorting. Science 238:323–329
57. Ibrahim SF, van den Engh G (2004) High-speed chromosome sorting. Chromosome Res 12(1):5–14
58. Liu AY, True LD, LaTray L, Nelson PS, Ellis WJ, Vessella RL, Lange PH, Hood L, van den Engh G (1997) Cell–cell interaction in prostate gene regulation and cytodifferentiation. Proc Natl Acad Sci USA 94:10705–10710
59. Visser JW, van den Engh GJ, van Bekkum DW (1980) Light scattering properties of murine hemopoietic cells. Blood Cells 6:391–407
60. Tamaki S, Eckert K, He D, Sutton R, Doshe M, Jain G, Tushinski R, Reitsma M, Harris B, Tsukamoto A, Gage F, Weissman I, Uchida N (2002) Engraftment of sorted/expanded human central nervous system stem cells from fetal brain. J Neurosci Res 69:976–986
61. Huntly BJ, Gilliland DG (2005) Leukaemia stem cells and the evolution of cancer-stem-cell research. Nat Rev Cancer 5(4):311–321
62. Bryant Z, Subrahmanyan L, Tworoger M, LaTray L, Liu CR, Li MJ, van den Engh G, Ruohola-Baker H (1999) Characterization of differentially expressed genes in purified

Drosophila follicle cells: toward a general strategy for cell type-specific developmental analysis. Proc Natl Acad Sci USA 96:5559–5564
63. Li J, Greco V, Guasch G, Fuchs E, Mombaerts P (2007) Mice cloned from skin cells. Proc Natl Acad Sci USA 104(8):2738–2743
64. Ibrahim SF, Diercks AH, Petersen TW, van den Engh G (2007) Kinetic analyses as a critical parameter in defining the side population (SP) phenotype. Exp Cell Res 313(9):1921–1926
65. Petersen TW, Ibrahim SF, Diercks AH, van den Engh G (2004) Chromatic shifts in the fluorescence emitted by murine thymocytes stained with Hoechst 33342. Cytometry A 60(2):173–181
66. Chen W, Georgiou G (2002) Cell-surface display of heterologous proteins: from high-throughput screening to environmental applications. Biotechnol Bioeng 79:496–503
67. Daugherty PS, Iverson BL, Georgiou G (2000) Flow cytometric screening of cell-based libraries. J Immunol Methods 243:211–227
68. Chalfie M, Tu Y, Euskirchen G, Ward WW, Prasher DC (1994) Green fluorescent protein as a marker for gene expression. Science 263:802–805
69. Misteli T, Spector DL (1997) Applications of the green fluorescent protein in cell biology and biotechnology. Nat Biotechnol 15:961–964
70. Matz MV, Fradkov AF, Labas YA, Savitsky AP, Zaraisky AG, Markelov ML, Lukyanov SA (1999) Fluorescent proteins from nonbioluminescent Anthozoa species. Nat Biotechnol 17:969–973
71. Arun KH, Kaul CL, Ramarao P (2005) Green fluorescent proteins in receptor research: an emerging tool for drug discovery. J Pharmacol Toxicol Meth 51(1):1–23
72. Chen G, Hayhurst A, Thomas JG, Harvey BR, Iverson BL, Georgiou G (2001) Isolation of high-affinity ligand-binding proteins by periplasmic expression with cytometric screening (PECS). Nat Biotechnol 19:537–542
73. VanAntwerp JJ, Wittrup KD (2000) Fine affinity discrimination by yeast surface display and flow cytometry. Biotechnol Prog 16:31–37
74. Boder ET, Midelfort KS, Wittrup KD (2000) Directed evolution of antibody fragments with monovalent femtomolar antigen-binding affinity. Proc Natl Acad Sci USA 97:10701–10705
75. Olsen MJ, Stephens D, Griffiths D, Daugherty P, Georgiou G, Iverson BL (2000) Function-based isolation of novel enzymes from a large library. Nat Biotechnol 18:1071–1074
76. Barker LP, Brooks DM, Small PL (1998) The identification of Mycobacterium marinum genes differentially expressed in macrophage phagosomes using promoter fusions to green fluorescent protein. Mol Microbiol 29:1167–1177
77. Waller A, Simons P, Prossnitz ER, Edwards BS, Sklar LA (2003) High throughput screening of G-protein coupled receptors via flow cytometry. Comb Chem High Throughput Screen 6(4):389–397
78. Krutzik PO, Nolan GP (2006) Fluorescent cell barcoding in flow cytometry allows high-throughput drug screening and signaling profiling. Nat Methods 3(5):361–368
79. Farinas ET (2006) Fluorescence activated cell sorting for enzymatic activity. Comb Chem High Throughput Screen 9(4):321–328
80. Krutzik PO, Hale MB, Nolan GP (2005) Characterization of the murine immunological signaling network with phosphospecific flow cytometry. J Immunol 175(4):2366–2373
81. Galbraith DW, Elumalai R, Gong FC (2004) Integrative flow cytometric and microarray approaches for use in transcriptional profiling. Methods Mol Biol 263:259–280

82. Choe J, Guo HH, van den Engh G (2005) A dual-fluorescence reporter system for high-throughput clone characterization and selection by cell sorting. Nucleic Acids Res 33(5):e49
83. Tay ST, Ivanov V, Yi S, Zhuang WQ, Tay JH (2002) Presence of anaerobic bacteroides in aerobically grown microbial granules. Microb Ecol 44(3):278–285
84. Perez OD, Nolan GP (2002) Simultaneous measurement of multiple active kinase states using polychromatic flow cytometry. Nat Biotechnol 20:155–162
85. Vermes I, Haanen C, Reutelingsperger C (2000) Flow cytometry of apoptotic cell death. J Immunol Meth 243:167–190
86. Pala P, Hussell T, Openshaw PJ (2000) Flow cytometric measurement of intracellular cytokines. J Immunol Methods 243:107–124
87. Lehmann AK, Sornes S, Halstensen A (2000) Phagocytosis: measurement by flow cytometry. J Immunol Methods 243:229–242
88. Stauber RH (2001) Methods and assays to investigate nuclear export. Curr Top Microbiol Immunol 259:119–128
89. Thiel A, Scheffold A, Radbruch A (2004) Antigen-specific cytometry – new tools arrived! Clin Immunol 111(2):155–161
90. Becker S, Schmoldt HU, Adams TM, Wilhelm S, Kolmar H (2004) Ultra-high-throughput screening based on cell-surface display and fluorescence-activated cell sorting for the identification of novel biocatalysts. Curr Opin Biotechnol 15(4):323–329
91. Eisenstein M (2006) Cell sorting: divide and conquer. Nature 441(7097):1179–1185
92. Leary JF (2005) Ultra high-speed sorting. Cytometry A 67(2):76–85
93. Shapiro HM (2004) Cellular astronomy – a foreseeable future in cytometry. Cytometry A 60(2):115–124

Adv Biochem Engin/Biotechnol (2007) 106: 41–73
DOI 10.1007/10_2007_072

Published online: 7 August 2007

Cell Isolation and Expansion Using Dynabeads®

Axl A. Neurauter (✉) · Mark Bonyhadi · Eli Lien · Lars Nøkleby · Erik Ruud · Stephanie Camacho · Tanja Aarvak

Invitrogen Dynal AS, Oslo, Norway
axl.neurauter@invitrogen.com

Abstract This chapter describes the use of Dynabeads for cell isolation and expansion. Dynabeads are uniform polystyrene spherical beads that have been made magnetisable and superparamagnetic, meaning they are only magnetic in a magnetic field. Due to this property, the beads can easily be resuspended when the magnetic field is removed. The invention of Dynabeads made, by Professor John Ugelstad, has revolutionized the separation of many biological materials. For example, the attachment of target-specific antibodies to the surface of the beads allows capture and isolation of intact cells directly from a complex suspension such as blood. This is all accomplished under the influence of a simple magnetic field without the need for column separation techniques or centrifugation.

In general, magnetic beads coated with specific antibodies can be used either for isolation or depletion of various cell types. Positive or negative cell isolation can be performed depending on the nature of the starting sample, the cell surface markers and the downstream application in question. Positive cell isolation is the method of choice for unprocessed samples, such as whole blood, and for downstream molecular applications. Positive cell isolation can also be used for any downstream application after detachment and removal of the beads. Negative cell isolation is the method of choice when it is critical that cells of interest remain untouched, i.e., no antibodies have been bound to any cell surface markers on the cells of interest. Some cell populations can only be defined by multiple cell surface markers. Such populations of cells can be isolated by the combination of negative and positive cell isolation. By coupling Dynabeads with antibodies directed against cell surface activation molecules, the beads can be used both for isolation and expansion of the cells.

Dynabeads are currently used in two major clinical applications: 1) In the Isolex® 300i Magnetic Cell Selection System for CD34 Stem Cell Isolation – 2) For ex vivo T cell isolation and expansion using Dynabeads® *ClinExVivo*™ CD3/CD28 for clinical trials in novel adoptive immunotherapy.

Keywords Cell expansion · Cell isolation · Dynabeads · Flow compatibility · Negative isolation · Positive isolation

1 Introduction

The characterization of specific cell types and the investigation of their functions, requires that the cells of interest can be isolated or purified from other contaminating cells. There are a number of isolation methods available for the specific isolation of cells using non-magnetic and magnetic separation. Flow-assisted cell sorting (FACS) is a non-magnetic method to obtain highly purified cells. This method, however, is quite time consuming, costly and can be rough on cells. Cell separation techniques based on the use of antibody-coated magnetic beads [1,2], are now widely used in research and clinical laboratories.

There are two types of magnetic cell isolation technologies, column-based and tube-based systems. The column-based technology utilizes smaller, nano-sized particles and therefore requires that the cells be passed through a magnetized iron-mesh column to increase cell-capture capacity. This technology will not be discussed further in this chapter.

The Dynabeads® tube-based system utilizes larger, micron-sized beads. Specific cells can, after binding to the antibody coated magnetic beads, be selected by the use of just a magnet which is held against the sample tube. Following brief washing, high cell purity can be achieved. Dynabeads® were invented by Professor John Ugelstad in 1977, and were the first monodispersed particles of their kind (Fig. 1). In 1976, after learning about plans by US scientists to manufacture monosized polymer particles in space under non-gravitational conditions, and enticed by the idea of producing such beads under normal conditions, Professor Ugelstad began work on developing monodisperesed beads. One year later, the first batch of such beads was manufactured in his lab at the Norwegian Technical University (NTH), and not until 5 years later, did the American researchers succeed in their quest to produce 10 g of monodispersed particles in a satellite orbiting Earth. By this time, Professor Ugelstad's monodispersed beads had already been mag-

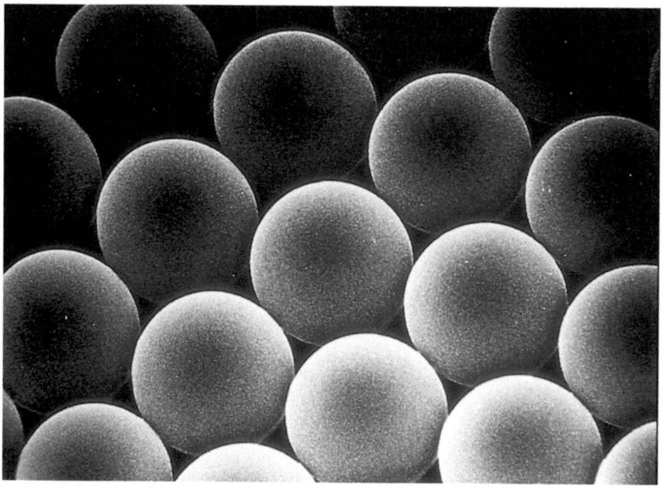

Fig. 1 Picture of Dynabeads: In 1976 the late Professor John Ugelstad of Norway achieved something otherwise only achieved by NASA in the weightless conditions of space: the making of uniform polystyrene spherical beads of exactly the same size. Professor Ugelstad and his colleagues made these beads magnetisable and superparamagnetic, meaning they are only magnetic in a magnetic field. Due to this property, the beads can easily be resuspended when the magnetic field is removed. This innovation revolutionized the separation of many biological material. For example, the attachment of target-specific antibodies to the surface of the beads allows capture and isolation of intact cells directly from a complex suspension such as blood. This is all accomplished under the influence of a simple magnetic field without the need for column separation techniques or centrifugation [1, 2]

netized and the technology patented. Shortly thereafter, the first products for biomagnetic separation became commercially available under the trademark Dynabeads®.

There are two main strategies for isolating a specific cell type: "positive isolation" of the cell of interest or "negative isolation" where unwanted cells are depleted (Fig. 2). By positive isolation, a specific cellular subset is isolated directly from a complex mixture of cells based on the expression of a distinct surface antigen. The resulting immune complexes of antibody coated beads and target cells are collected using a magnet. For functional studies, the beads should be removed from the positively isolated cells. There are a number of ways in which Dynabeads can be released from the cells after isolation using competing molecules that through affinity binding competition release the bead or antibody-bead complex from the cells of interest.

By negative isolation, all unwanted cell types sensitized with antibodies are removed from the sample by the magnetic beads. Cells isolated by negative isolation have not been bound to antibodies at any time. This is an advantage, as surface antigen-bound antibodies may elicit the transmission of signals across the cell membrane.

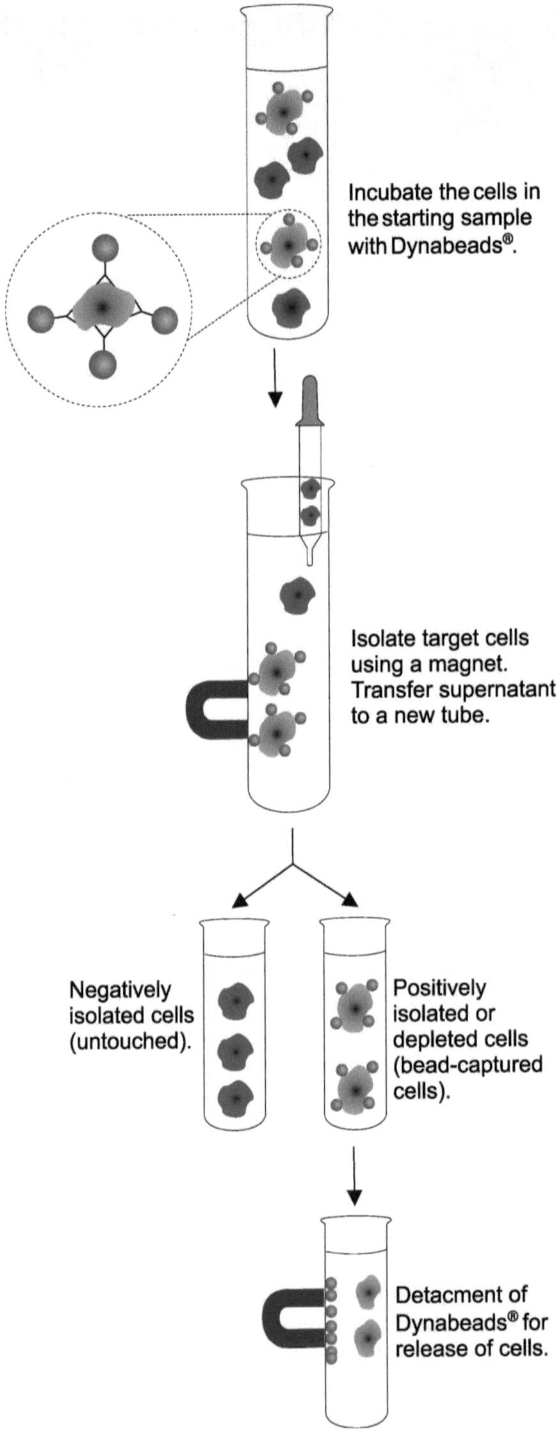
Incubate the cells in the starting sample with Dynabeads®.
Isolate target cells using a magnet. Transfer supernatant to a new tube.
Negatively isolated cells (untouched).
Positively isolated or depleted cells (bead-captured cells).
Detacment of Dynabeads® for release of cells.

◀ **Fig. 2** Schematic illustration of positively and negatively isolated cells using Dynabeads

Factors such as incubation time, temperature, and concentration of reactants have a measurable effect on the efficiency of cell isolation using magnetic beads. Furthermore, the process is also affected by specific parameters, such as the nature and state of the target cell, characteristics of the antigen/antibody binding, sample type, concentration, and ratio of beads and cells. Successful cell isolation with Dynabeads, which implies high yield and purity, is dependent on the concentration of the magnetic beads, the ratio of beads to target cells, and the choice of antibody. Monoclonal antibodies are generally recommended due to their high specificity towards the target antigen.

The immunomagnetic isolation technique continues to encompass new fields for the selective isolation of eukaryotic cells [3–7], and the use of pure cell populations has now reached the field of therapy through the initiation of clinical trials. Ex vivo isolation and expansion of cells have given promising possibilities in immunotherapy. Dynabeads coated with signaling molecules such as anti-CD3 and anti-CD28 antibodies have proven to be very efficient in the in vitro activation of T cells, the prime effectors of the acquired immune system [8,9]. An ex vivo-expanded population of T cells may be administrated to the patient, thereby helping to fight diseases such as cancer, HIV, and autoimmune disorders [10–13].

2 Positive Cell Isolation

Positive cell isolation is defined as the method whereby a single cell type is directly drawn out of a mixture of cells using cell-specific antibodies or ligands linked to magnetic beads. The antibodies or ligands can be covalently attached directly to the magnetic bead or – for greater flexibility – via secondary antibodies.

Due to the characteristics of Dynabeads, a wide variety of starting samples can be used for positive cell isolation. Highly pure cells can be isolated from non-processed samples such as whole blood, umbilical cord blood, bone marrow, feces and ascites, or from processed samples such as buffy coat, peripheral blood mononuclear cells (PBMC), tissue digests or cell cultures.

Depletion of cells, i.e., when a cell population bound to the magnetic beads via a primary antibody is removed from a cell sample, follow the same rules as positive cell isolation. The rest of the chapter will discuss positive cell isolation and depletion of cells as one.

Positive cell isolation using magnetic beads has over the last 20 years become a powerful tool to drive research within a wide range of life science fields such as cell biology, immunology, cancer research, in vitro

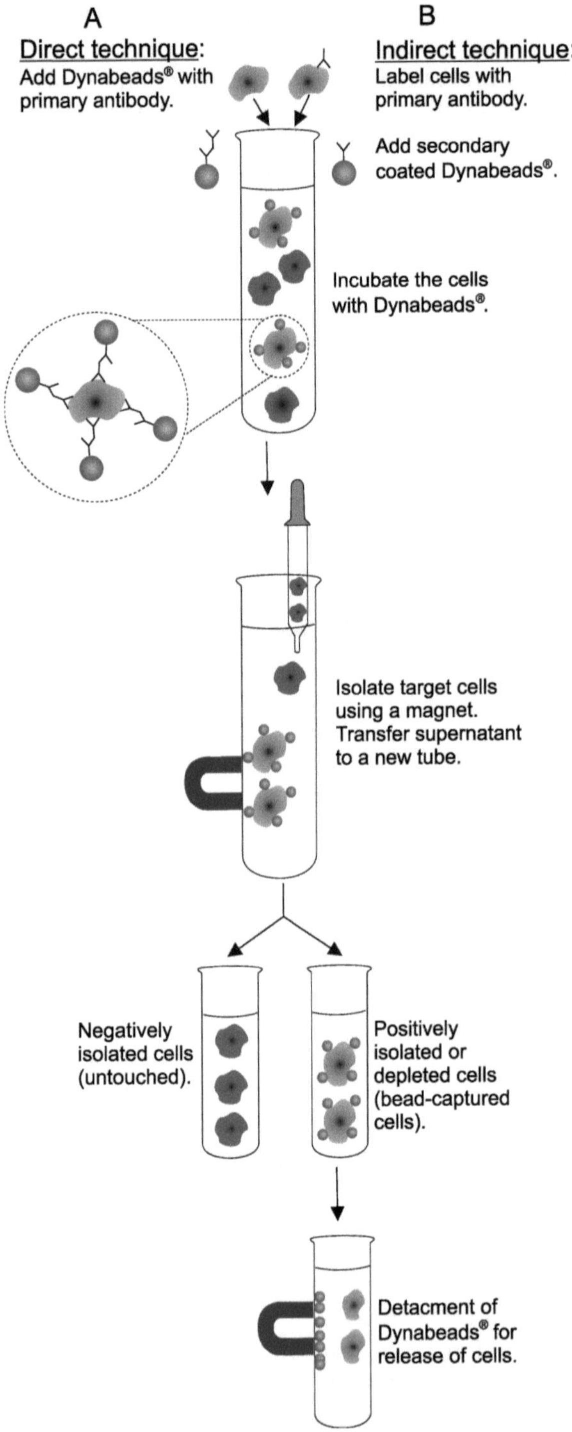
A
Direct technique:
Add Dynabeads® with primary antibody.
B
Indirect technique:
Label cells with primary antibody.
Add secondary coated Dynabeads®.
Incubate the cells with Dynabeads®.
Isolate target cells using a magnet. Transfer supernatant to a new tube.
Negatively isolated cells (untouched).
Positively isolated or depleted cells (bead-captured cells).
Detacment of Dynabeads® for release of cells.

◀ **Fig. 3** Schematic illustration of **A** direct and **B** indirect techniques for positively and negatively isolated cells

diagnostics and cell therapy (see Table 1). Crucial for this success has been 1) the availability of a wide range of antibodies [14] to cell differentiation molecules (e.g., human cell differentiation molecules (HCDM); http://www.hlda8.org/HLDAtoHCDM.htm), and 2) the result of the development of micron-sized, super-paramagnetic polymer beads, called Dynabeads [1]. The combination of monoclonal antibodies and magnetic beads allow quick (10–30 minutes) and gentle isolation of specific cells.

Positive cell isolation can be performed using two different approaches, i.e., the direct or indirect technique. Using the direct technique, one links the primary antibody or ligand to the beads (primary coated beads) prior to cell isolation. The primary coated beads are mixed with the cell sample, and after a short incubation period, whereby the cells bind to the primary coated beads, cells are easily collected with the aid of a magnet (Fig. 3A).

Using the indirect technique, one utilizes secondary coated Dynabeads. The cell sample is first mixed with the primary antibody (or other binding molecule, defining the target), and after a short incubation, excess antibody may be removed by centrifugation. Secondary coated Dynabeads (coated with

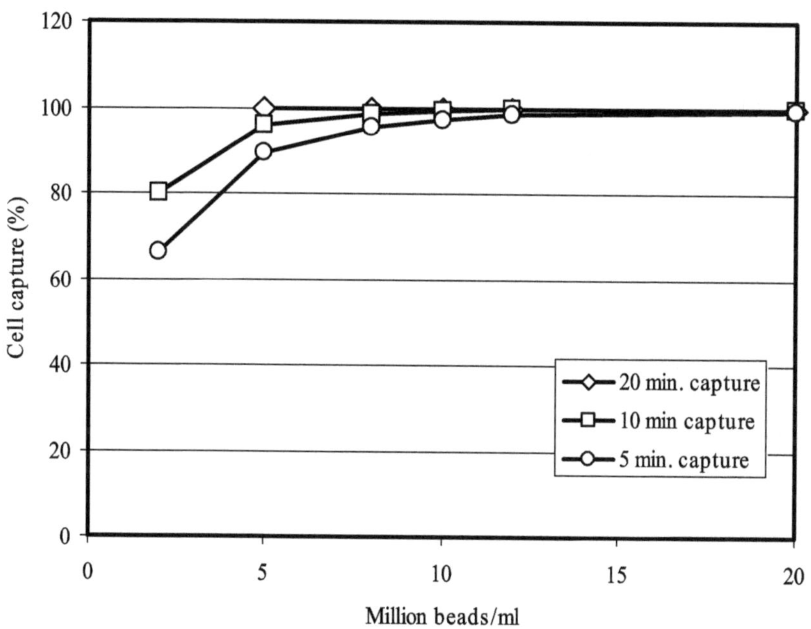

Fig. 4 Isolation of human CD3+ cells using primary coated Dynabeads and direct technique, coated with a high affinity antibody. Human CD3+ cells were isolated from whole blood using 2–20 million beads/ml, 5–20 minutes incubation time

an antibody or other molecules which can bind the primary antibody) are then mixed with the sample, and during a short incubation the cells of interest will bind to the beads, and can thereafter be isolated by the aid of a magnet (Fig. 3B).

The advantage of the direct isolation technique, is that it can be used with cell samples such as whole blood, especially when high affinity antibodies (or other binding molecules) are used (Fig. 4). The cell capture kinetics may, however, be slower in the direct technique compared to the indirect technique, especially if low affinity antibodies (or other binding molecules) are involved (Fig. 5).

Primary coated Dynabeads, used in the direct technique, are commercially available products designed to isolate specific cells or may be designed by the researchers themselves by coupling the primary antibody (or other binding molecule) onto surface activated Dynabeads and secondary coated Dynabeads. Primary coated Dynabeads have been used for isolation of a range of cells from several species (see Table 1).

Secondary coated Dynabeads, used both in the direct and indirect technique, are very flexible products designed to isolate any cell of interest. The limitation of this product range lies in the availability of appropriate target

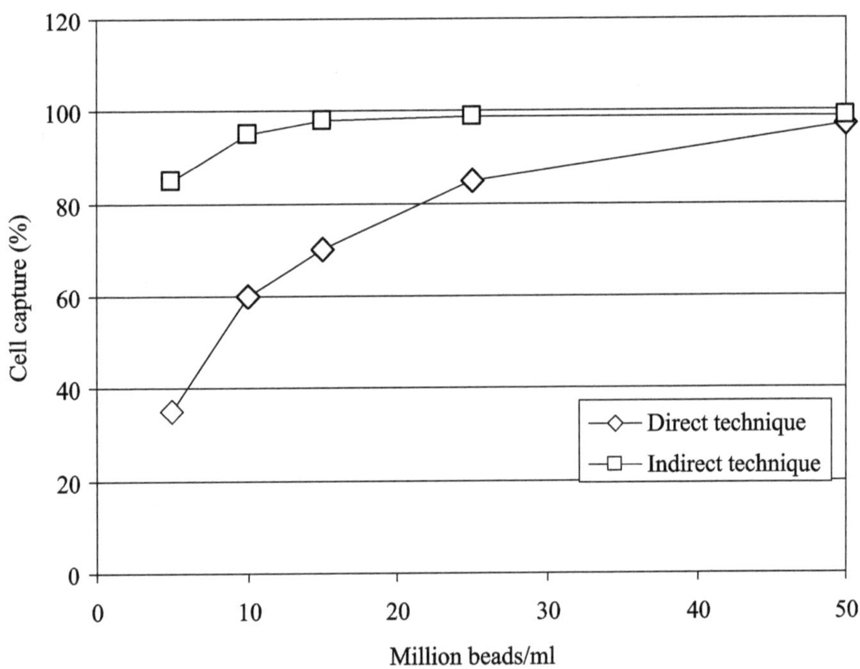

Fig. 5 Isolation of Daudi cells using secondary coated Dynabeads and a low affinity primary antibody. Daudi cells were isolated from cell culture medium using 5–50 million beads/ml in the *direct* and *indirect technique*, respectively, 10 minutes incubation time

defining binders, e.g., primary antibodies, super antigens, recombinant major histocompatibility complexes (recomb. MHC) or tags for cells surface antigens. Due to their flexibility, secondary-coated Dynabeads have been used for isolation of a wide range of cells from several species (Table 1).

The challenge when performing positive cell isolation is that the cell of interest may be affected and/or altered during the isolation step. In most systems, attached antibody coated beads remain on the cells throughout the duration of downstream experiments. The binding of antibodies or ligands to cell surface antigens can lead to clustering of receptors, triggering of signaling pathways (positive or negative signal) or blocking of receptor function. Therefore, choosing the right surface markers and corresponding antibody producing clones are critical. In addition it is important that the magnetic beads used for isolation are inert and non-degradable in order to prevent cell damage by iron oxide exposure to cells. [15–17]. Dynabeads have an even dispersion of superparamagnetic material coated with a polymer shell that encases the magnetic material to prevent iron oxide leakage.

Positive cell isolation without bead removal can be used for many downstream applications, such as molecular biological analysis including nucleic acid or protein analysis. However, in order to perform other downstream applications, i.e., flow cytometry and cell cultures, it is necessary to remove the beads due to their micron-size. Therefore various bead-release methods have been developed.

Table 1 Overview over various human and mouse cells isolated with Dynabeads

Cell type (positive/negative isolation or depletion)	Cell subset	Refs.
Human T cells	General	[70, 71]
	CD3+ T cells	[72, 73]
	CD4+ T cells	[74–81]
	CD8+ T cells	[76, 78–84]
	Naïve (CD45RA+)	[85]
	Memory (CD45RO+)	[85]
Human regulatory T cells (CD4+CD25+)		[86–89]
Human B cells (CD19+)		[90–96]
Human NK cells (CD56+)		[97–100]
Human NKT cells		[101]
Human monocytes (CD14+)		[102–104]
Human dendritic cells		[105–109]
Human progenitor cells (CD34+)		[110–115]
Human granulocytes	General	[116, 117]
	Neutrophils (alveolar)	[118]
	Eosinophils	[119–121]
	Basophils	[122]

Table 1 (continued)

Cell type (positive/negative isolation or depletion)	Cell subset	Refs.
Human megakaryocytes		[123]
Human platelets		[124]
Human mast cells		[125]
Human endothelial cells	General	[126, 127]
	Mammary adipose endothelial cells	[128]
	Hepatic endothelial cells	[129]
	Circulating endothelial cells	[130–132]
Human fibroblasts		[133, 134]
Human epithelial cells	General	[135, 136]
	Colorectal cancer cells	[137]
	Mammary carcinoma cells	[138]
Human spermatozoa		[139]
Human osteoclasts		[140]
Human HIV infected cells		[141]
Mouse T cells	General	[142–145]
	CD4+ T cells	[146–150]
	CD8+ T cells	[151–155]
Mouse regulatory T cells (CD4+CD25+)		[156]
Mouse B cells		[157, 158]
Mouse NK cells		[159]
Mouse dendritic cells		[160, 161]
Mouse endothelial cells	Myocardial endothelial cells	[162]
	Embryonic endothelial progenitor cells	[163]
	Skin endothelial cells	[164]
	Lung endothelial cells	[165, 166]
Mouse Langerhans cells		[167]
Mouse hematopoietic progenitor (Lin-) cells		[168–170]

2.1 Detachment of Dynabeads® Following Positive Cell Isolation

Various technologies have been developed to remove Dynabeads from cells. These can be divided into two categories: 1) Detachment of the Dynabeads and the primary antibody from the cell surface with antibody-specific release mechanisms and 2) Detachment of the Dynabeads alone with generic release mechanisms, i.e., enzymatic cleavage or affinity molecules.

DETACHaBEAD® is a commercially available technology whereby the cells are released from the antibody-beads. The DETACHaBEAD reagent is based on antibodies that were raised against the primary antibody that recognizes the cell surface marker. The anti-antibodies recognize the variable region of the primary antibody and thereby disrupt the binding between the primary antibody and the surface marker. As these anti-antibodies out-compete the binding between the primary antibody and the surface marker, cells are released, leaving the cells free of beads and antibodies [18–22].

Peptides can also be used to specifically out-compete the binding between the primary antibody on the bead and the surface marker of the cell. One such application has been developed by Baxter Healthcare Corporation, which has the only FDA- and CE mark-approved device in the field of stem cell therapy – the Isolex® 300i Magnetic Cell Selection System. The technology uses an anti-CD34 monoclonal antibody and Dynabeads to isolate CD34+ stem cells. The isolated cells are released using a peptide (PR34+™ releasing agent) resulting in antibody free CD34+ stem cells. Isolex® and PR34+™ are trademarks of Baxter International Inc. [23–27].

FlowComp™ is a new commercially available generic release technology for research use where the release is based on affinity-competition of a biotin analogue to an avidin/streptavidin analogue (Fig. 6). The advantage of this technology is the speed of release (5–10 minutes) and because it is generic it can be applied to most cell types (Fig. 7).

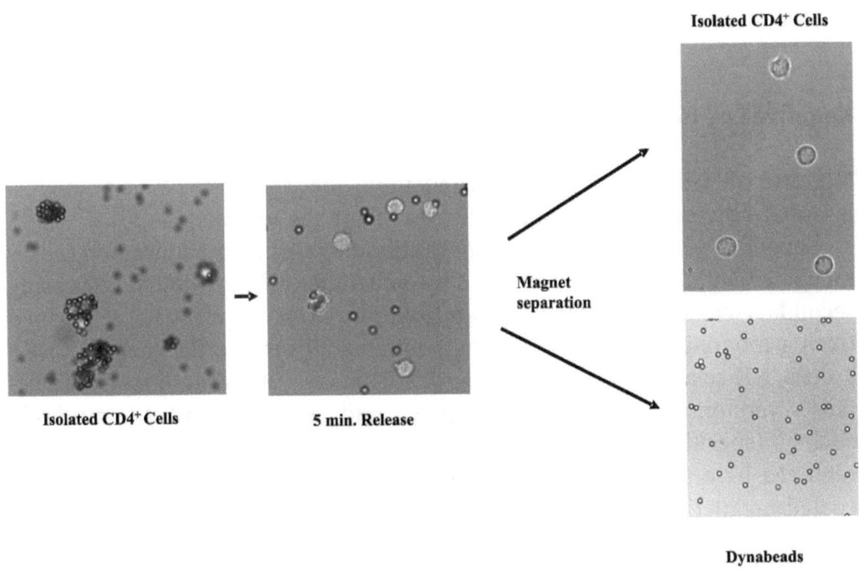

Fig. 6 Picture of CD4+ cells isolated and released by the FlowComp technology. PBMCs was isolated from a healthy donor, stained with Hoescht 33342 and CD4+ cells were isolated using Dynabeads FlowComp Human CD4. Cells were visualized using a fluorescence microscope after cell capture (Isolated CD4+), after 5 minutes incubation with FlowComp release buffer (5 min) and after magnetic separation of released cells (Isolated CD4+ and Dynabeads, respectively)

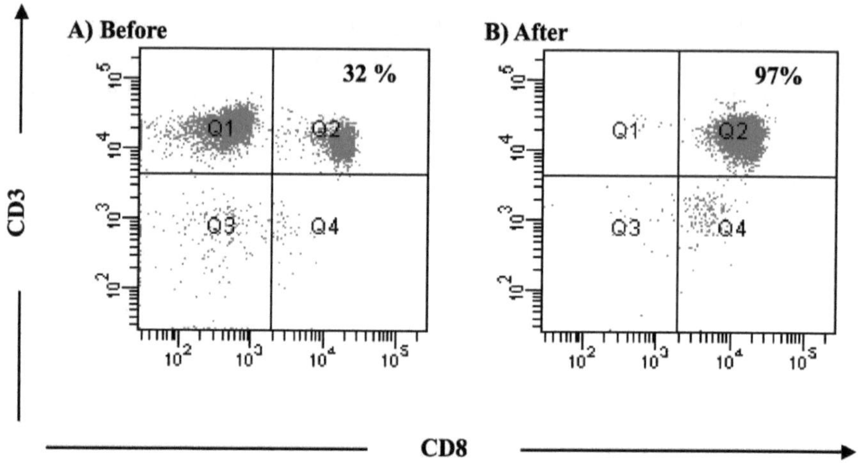

Fig. 7 CD8+ cells before and after isolation with the FlowComp technology. PBMC was isolated from a healthy donor, stained with CD8-FITC vs. CD3-PE before (**A**) and after (**B**) isolation using Dynabeads FlowComp Human CD8. Histogram A was gated on live cells of the lymphocyte population, giving 32% CD3+CD8+ T cells (Q2 in histogram A). Histogram B was gated on live cells of the lymphocyte + monocyte population, giving 97% purity of CD3+CD8+ T cells (Q2 in histogram B). Some $CD3^{-}CD8^{dim}$ cells (NK-cells) are co-isolated using this method (Q4 in histogram B)

3 Negative Cell Isolation of Untouched Cells

Negative cell isolation is defined as a method whereby the cell type of interest is isolated by removing all other cell types from the sample [28–30]

Generally a cocktail of monoclonal antibodies against the unwanted cells is incubated with the sample, followed by depletion of the unwanted cells using secondary-coated Dynabeads. As an example, CD4+ T cells can be purified from a mononuclear cell sample by removing all other cells, such as CD8+ T cells, NK cells, B cells, monocytes, dendritic cells and granulocytes. In details – mononucleated cells are prepared and incubated with the antibody cocktail (specific for non-CD4 cells) for 10–20 minutes. Excess antibodies are removed by a short centrifugation step, followed by addition of secondary coated magnetic beads. During a 15-minute incubation the unwanted, antibody sensitized cells, will bind to secondary coated magnetic beads and then be removed by the aid of a magnet.

Negative cell isolation can be used both for pre-enrichment of rare cells (e.g., dendritic cells) (Fig. 8), or isolation of a highly pure cell population (e.g., CD4+ T cells) (Fig. 9).

The advantage of negative cell isolation is that the cells of interest have not been attached to the antibodies on the magnetic beads at any time, and

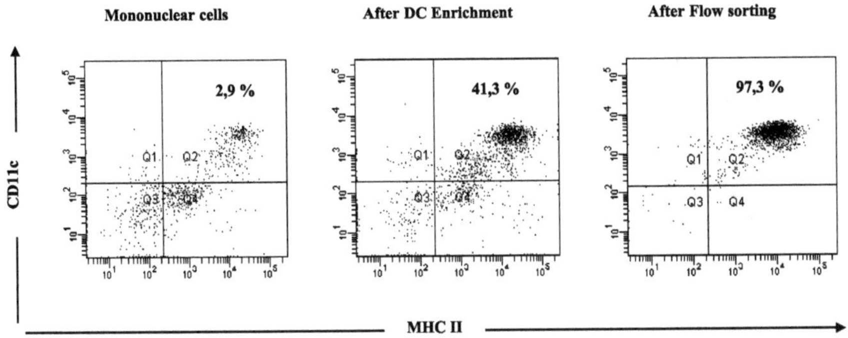

Fig. 8 Purity of mouse Lin-CD11c+MHCII+ DC after enrichment and flow sorting. Mouse spleen mononuclear cells were isolated from Balb/C, stained with Lin-PE cocktail vs. CD11c-FITC vs. MHC II-APC before isolation, after enrichment and after flow sorting. Enrichment was done using "Dynabeads Mouse DC Enrichment Kit", and flow sorting was done on the Dynabeads-enriched population. All histograms were gated on live cells

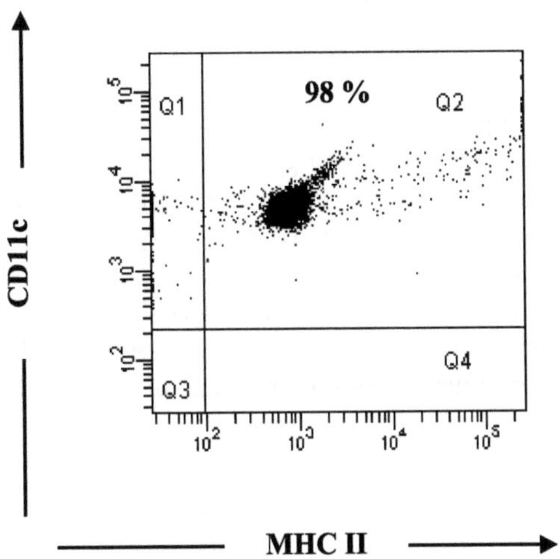

Fig. 9 Highly pure CD4+ T cells after negative isolation. PBMC was isolated from a healthy donor and CD4+ T cells isolated using Dynabeads MyPure CD4 Kit 2. Cells were after isolation stained for CD4-FITC vs. CD45-PE. The histogram was gated on live cells of the lymphocyte + monocyte population in the forward scatter (FSC) vs. side scatter (SSC) histogram, giving 98% purity of CD45+CD4+ T cells (Q2 in histogram)

thereby avoiding any possible antibody induced signaling through cell surface molecules.

The isolated cells are ideal for various applications in HIV, cancer and autoimmune disease research, such as flow cytometric analyses, signalling pathway and gene transcription experiments, cytokine secretion assays, cytotoxicity studies, cell culture etc.

4 Isolation of Subpopulations of Cells – Combination of Negative and Positive Isolation

Major cell populations defined by specific cell surface markers (e.g., CD3+ T cells, CD19+ B cells, CD14+ monocytes) can be isolated with high purity using either positive or negative cell isolation (described above). However, there exist subpopulations of cells that are not defined by any specific marker, and therefore combinations of several markers must be used for their isolation.

The way to achieve a minor subpopulation is to use a combination of negative and positive cell isolation. A combination strategy can also be used for minor cell populations to achieve an improved purity by first depleting a fraction of the unwanted cell population before a positive isolation of the wanted cell population.

As an example, regulatory T cells (Treg cells) are a subpopulation of T cells defined by co-expression of the CD4 and CD25 surface markers. The Treg cell population constitutes 2–10% of the total CD4+ T cells in peripheral blood. Treg cells are specialized for suppression of immune responses and play a critical role in maintaining immunological self-tolerance by actively suppressing self-reactive lymphocytes. Since a single surface marker for the identification of regulatory T cells has to date not yet been identified, these cells must be isolated using a two-step isolation strategy; initial negative cell isolation of CD4+ cells followed by positive cell isolation of the CD25+ Treg cells [31, 32].

Several of the surface markers expressed by Treg cells (Fig. 10) are also expressed by conventional T cells [32]. CD25 is upregulated on activated conventional cells but is constitutively expressed on Treg. In human peripheral

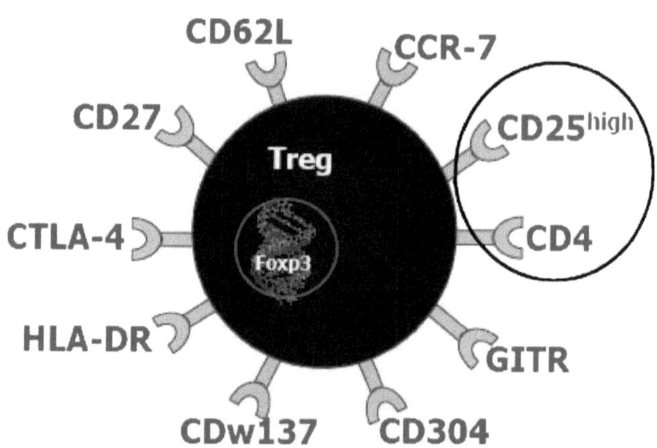

Fig. 10 Picture of surface markers known to be expressed by regulatory T cells. The co-expression of CD4 and CD25 defines the Treg cell

blood the majority of the T cells are of naive phenotype and therefore a combination of CD4 and CD25 can be used to isolate a highly pure population of Treg cells.

The principle of isolation is illustrated in (Fig. 11). First negative cell isolation of CD4+ T cells is performed by depleting the CD8,CD14, CD16, CD19, CD36, CD56,CDw123 and CD235a from mononuclear cells. Following isolation of the CD4+ cells, Dynabeads® CD25 are added to positively isolate the regulatory CD4+CD25+ T cells. After positive isolation of the CD4+CD25+ Treg cells, the cells are detached from the beads by adding DETACHaBEAD reagent.

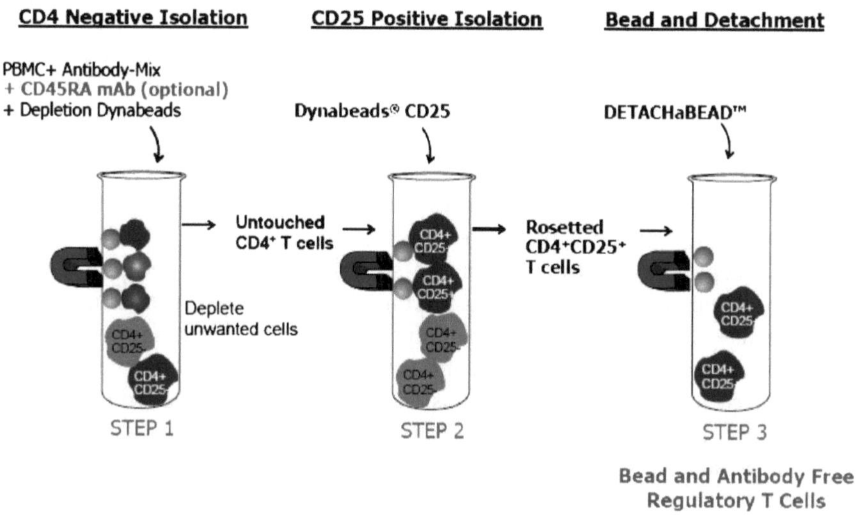

Fig. 11 Schematic illustration of how Treg cells can be isolated using a combination of negative and then positive isolation. DETACaBEAD is used after positive isolation to release the beads from the cells

By using this approach regulatory T cells can be isolated with a purity of 95–98% [31]. Figure 12 shows the presence of CD4+CD25+ cells during the various isolation steps. In PBMC sample, 1–3% of the cells are CD4+ CD25+. After a CD4+ negative isolation step the purity of CD4+ cells is 95–98%, and 3–8% of these cells express CD25. When the CD25+ cells are positively isolated with the Dynabeads CD25 the depleted fraction of CD4+ cells contains only 0.5–1.5% CD25+ cells. The resulting CD4+CD25+ subpopulation is 95–98% pure.

If the number of cells in a subpopulation of interest is low, as for the CD4+CD25+ Treg population, an expansion step can be performed to obtain a sufficient number of cells needed to perform downstream functional assays. Both conventional T cells and regulatory T cells can be expanded 100–1000-fold when activated properly [33]. In vivo, T cells are activated

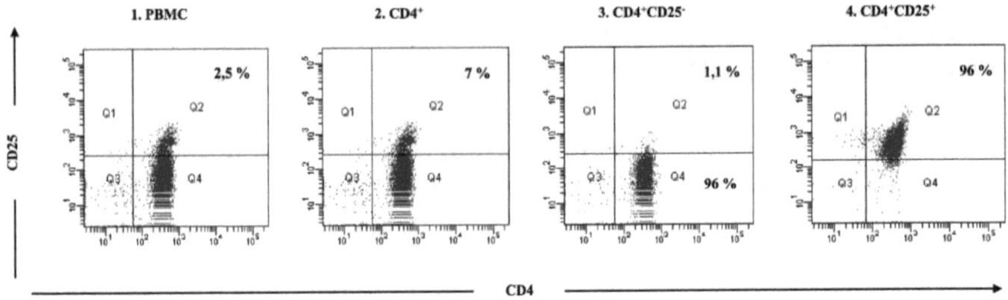

Fig. 12 The presence and percentage of CD4+CD25+ T cells during the various isolation step from 1) PBMC, 2) CD4+ cells, 3) depleted CD4+CD25- and 4) the isolated CD4+CD25+ cells. PBMCs was isolated from a healthy donor and stained with CD4-PerCP vs. CD25-PE before (1) and after (2) isolation using Dynabeads MyPure CD4 Kit 2. Within the isolated CD4+ T cell population, 7% of the CD4+ cells expressed the CD25-marker. Further, Dynabeads CD25 were added to the CD4+ T cells to positively isolate CD4+CD25+ Treg cells, and cells were released from the beads using DETACHaBEAD. After isolation of CD25+ cells, 1% of the CD25+ cells (CD25^{dim}) were left in the CD25-depleted fraction (3), and the isolated CD25+ cells were 96% pure (4)

when they interact with antigen presenting cells (APCs) such as dendritic cells (DC). DC activate T cells after interaction with the CD3/TCR (TCR = T Cell Receptor) receptors and the co-stimulatory surface receptor CD28. In vitro, T cells can be activated using an artificial antigen presenting cells, Dynabeads® CD3/CD28 that are coated with affinity-purified monoclonal antibodies directed against CD3/TCR and the co-stimulatory CD28 receptor that are required for optimal T cell expansion. The beads are designed to expand T cells in a manner that mimics what occurs in vivo upon activation via antigen-presenting cells (Fig. 13). The Dynabeads expansion method elimi-

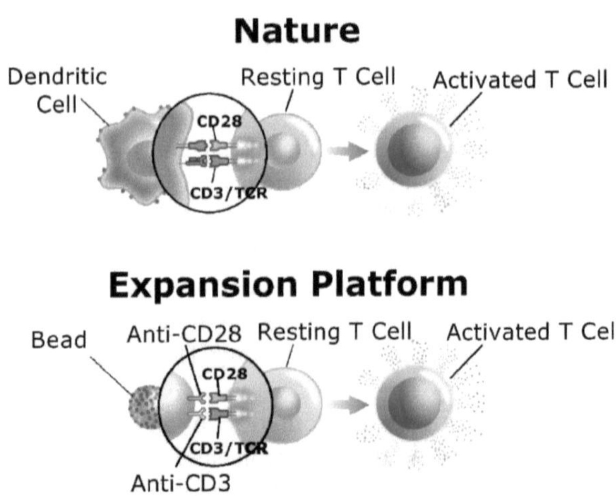

Fig. 13 Dynabeads CD3/CD28 mimics antigen presenting cells and activates T cells after interaction with the TCR/CD3 receptor and the co-stimulatory receptor CD28 on the T cell

nates the need to maintain autologous antigen-presenting cells and antigen in culture, making it the most reproducible and reliable way to stimulate T cells. Covalent attachment of antibodies to beads allows for easy magnetic removal of beads and antibodies after T cell expansion. Expanded Treg cells can also be used for down stream applications such as pre-clinical studies.

5 Cell Isolation and Expansion for Clinical Applications

Dynabeads have gained an important position in cell-based clinical applications due to their non-toxic nature, and the ability to manufacture antibody coated beads under sterile cGMP conditions. The scalability of the cell processing procedures enables handling of large cell numbers needed for clinical applications. As an example, Baxter Healthcare Corporation has adapted the Dynabeads technology in their Isolex® 300i Magnetic Cell Selection System for CD34 Stem Cell Isolation – the only FDA-approved and CE-marked technology for the clinical selection of CD34 positive cells, allowing the passive depletion of tumor cells from autologous apheresis products. The instrument has a capacity of up to 8×10^{10} nucleated cells and uses an anti-CD34 monoclonal antibody and Dynabeads for positive cell isolation of CD34+ stem cells ex vivo. The isolated stem cells are released using a peptide (PR34+™ releasing agent) resulting in antibody free cells [23–27]. Isolex® and PR34+™ are trademarks of Baxter International Inc.

5.1 Dynabeads® CD3/CD28 for T cell Isolation, Activation and Expansion

In the mid-1990s, professor Carl June's laboratory at University of Pennsylvania, Philadelphia, USA, found that by immobilizing anti-CD3 and anti-CD28 antibodies on Dynabeads, they could support the long-term proliferation of T cell in culture, while maintaining desirable biological characteristics [34–37]. Conventional approaches that used feeder cells for activation of T cells are cumbersome, impractical and prone to engage receptors which inhibit T cell growth. Additionally, in the clinical setting, conventional feeder cells are difficult to maintain at sufficient numbers and represent significant regulatory challenges. Moreover, the covalent attachment of the anti-CD3 and anti-CD28 antibodies on Dynabeads allows for easy magnetic removal of beads and antibodies prior to infusion of T cells into patients.

Dynabeads® *ClinExVivo*™ CD3/CD28 are designed for isolation [38,39], activation [40] and expansion [41,42] of human T cells. Firstly, CD3+ T cells are separated and concentrated from apheresis product by positive cell isolation using Dynabeads *ClinExVivo* CD3/CD28 capture via the anti-CD3 antibody. The isolation step is crucial to remove contaminating cell types which can influence the following activation and expansion steps [38, 39]. Alternative isolation strategies of CD3primary-coated Dynabeads T cells, such as elutriation [43] combined with magnetic antibody coated Dynabeads depletion of B cells can also be used.

Such alternative isolation methods will allow greater flexibility in manipulation of the bead to T cell ratio during activation and expansion with Dynabeads *ClinExVivo* CD3/CD28, which in turn can be beneficial for activating various subpopulations of T cells, e.g., antigen specific T cells [44, 45].

Secondly, T cells receive activation signals from Dynabeads *ClinExVivo* CD3/CD28 through TCR/CD3 complex and the co-stimulatory CD28 receptor that are required for optimal T cell activation [36]. Furthermore, T cells can be expanded 100–1000 fold using Dynabeads *ClinExVivo* CD3/CD28 if maintained in IL-2 containing media [42].

Both research and clinical-scale protocols have been developed for expanding T cells (Xcellerated T Cells™) in a variety of disease settings. Moreover, clinical-scale cGMP protocols using Dynabeads *ClinExVivo* CD3/CD28 have been developed that utilize cost and labor-saving bioreactor systems capable of reproducibly generating $> 1 \times 10^{11}$ T cells in a single culture bag/reactor in less than two weeks [42].

5.2 Characteristic of T cells Activated and Expanded with Dynabeads® CD3/CD28 and Dynabeads® *ClinExVivo*™ CD3/CD28

Bead-activated T cells exhibit a number of desirable qualities. T cell receptor (TCR) repertoire is maintained during the expansion process for polyclonal T cells, thereby preserving the broadest Ag-recognition capabilities [35, 40, 46, 47]. Surface CD28 expression is maintained, while key homing receptors (e.g., L-selectin) and survival molecules (Bcl-XL) are induced, thereby enhancing in vivo survival and homing potential [41, 46, 48]. Both CD4+ and CD8+ T cells are expanded by the process, thereby preserving both cytolytic/killing and T helper functions [46]. During the activation process, T cells are induced to express or secrete a wide array of key immunomodulatory molecules, including surface-bound CD40-ligand, CD137 (4-1BB), and cytokines, such as IL-2, IFNγ, and TNFα [35, 38, 42]. Importantly, in settings where T cells are typically found to be anergic or non-responsive, activation and expansion using the bead technology can reverse anergy and restore responses to mitogenic or specific antigen-stimulation [49–51]. The bead activation process can be further manipulated to control the expansion of antigen specific memory T cells [44]. Also, in the setting of regulatory T cells, it has been demonstrated that bead-expanded regulatory T cells exhibit potent inhibitory activity [52–55]. Finally, bead-activated T cells are easy to transduce with standard Moloney-based and lentivirus-based retroviral gene-transduction systems [47, 56].

5.3
Current and Future Applications

Pre-clinical Research Applications

Dynabeads CD3/CD28 and Dynabeads *ClinExVivo* CD3/CD28 have been and continue to be used extensively in preclinical studies evaluating the use of novel adoptive T cell transfer approaches for treating cancer, infectious disease, autoimmunity, and disorders associated with chemotherapy and allogeneic transplantation. T cells have been effectively expanded from cancer patients, chronic lymphocytic leukemia [CLL] [38], multiple myeloma [MM] [57], non-Hodgkin's lymphoma [NHL] [50, 58], renal cell carcinoma [RCC] [59], prostate cancer [PC] [60], breast cancer [BC] [45], HIV-infected patients [61, 62], and patients with autoimmune disease [63–65].

Clinical Applications

The safety of bead-activated T cells has or is being evaluated in over a dozen phase I clinical studies, including:

- Polyclonal CD4+ T cells for treating HIV infection [49]
- Gene-modified T cells for treating HIV infection [48, 61]
- Autologous polyclonally activated T cells for treating RCC, NHL, CML, PC, MM, CLL [50, 59, 67]
- Allogeneic donor lymphocyte infusion in AML, ALL, CLL, NHL [58, 66]
- Restoration of immunity in lymphopenic MM patients by vaccination and adoptive T-cell transfer [67]

In addition to the safety profile of the infused T cells observed in these trials, a number of other observations are noteworthy. In stem cell transplant settings with concurrent chemotherapy-induced lymphodepletion, the infusion of polyclonal bead-activated and expanded T cells after transplant resulted in the early recovery of both CD4+ and CD8+ T cells. In the setting of CLL, following infusion of bead-activated and expanded autologous T cells, a majority of patients experienced a significant reduction in both lympadenopathy [68] and splenomegaly [69]. Finally, in a number of clinical trials, infused T cells were long lived and elevated T cell counts following infusion were maintained for extended periods of time.

6
Comments for Practical Approach

Immunomagnetic cell isolation offers rapid and direct access to target cells from whole blood and bone marrow without cell loss or damage. If necessary, unwanted elements such as erythrocytes, free DNA, fat, or serum proteins may be removed before cell isolation to improve the performance of the beads. Buffy coat, a concentrate of white blood cells, has the advantage of high target cell concentration. However, the quality of buffy coat preparations may vary considerably. Density gradient isolation of cells provides removal of possible interfering elements and the ability to manipulate target cell concentration to perform cost-efficient cell isolation. Drawbacks include cell losses during centrifugations and negative effects on the cells due to contact with the density gradient medium. Generally the concentration of nucleated cells should be $\leq 5 \times 10^7$ cells/ml when performing immunomagnetic cell isolation. By positive isolation the cells of interest are isolated for analysis. General isolation parameters are $\geq 1 \times 10^7$ beads/ml, bead: cell ratio 4 : 1–10 : 1, and 10–30 min of incubation. typically, 95–100% purity and viability are achieved with 60–95% yield. For some downstream applications the beads can remain attached to the cells (e.g., mRNA or DNA isolation). By negative isolation (= depletion), unwanted cells are removed prior to analysis of the remaining population. General isolation parameters are $\geq 2 \times 10^7$ beads/ml, bead: cell ratio $\geq 4:1$, and 20–60 min of incubation. Typically 95–99% depletion of unwanted cells is achieved. Two successive depletion cycles may result in higher purity for small cell populations. The direct technique offers fast cell isolation with antibody-coated beads. Cell handling is minimized, reducing the risk for cell damage and loss. The indirect technique is especially useful when the affinity/avidity of the primary antibody is low or when the epitope density on the target cell is limited. The disadvantage of the indirect technique is cell handling (centrifugation). Cross-reactivity and Fc binding can be blocked by the addition of free proteins (e.g., Fc receptor blocking with γ-globulin). Nonspecific binding of "sticky" cells to beads can be avoided by pre-depletion of these cells with protein-coated beads (e.g., secondary coated or BSA-coated beads). Free DNA will contribute to binding of unwanted cells to the beads. Freezing/thawing may damage cells and release free DNA. DNase treatment prior to cell isolation will abolish this problem. Platelets may also induce nonspecific binding of cells to Dynabeads. Preparation of PBMC with a low platelet content will reduce this problem. Primary coated Dynabeads are ready-to-use products for a wide variety of cell surface markers. In addition, secondary coated Dynabeads offer an excellent possibility to make beads with the reactivity of choice using mouse, rat, or rabbit antibodies directly from culture supernatant, ascites, or polyclonal sera (without the need of purification). This is especially useful when only small amounts of nonpurified antibody are available. However, affinity-purified antibodies are preferred.

6.1 Pitfalls

1. Prolonged incubation (> 60 min) and increased bead concentration (> 1×10^8 beads/ml) will rarely improve the cell isolation efficiency. However, nonspecific binding may increase, damage of cells from sheer forces of the beads may occur, and risk of cell trapping increases.
2. A soluble form of cell surface antigens or other serum components can reduce the efficiency of immunomagnetic cell isolation. One or two washing steps will overcome this problem.
3. Non-specific binding. Genomic DNA from lysed cells (e.g., present in buffy coat, PBMC, or after freezing/thawing of cells) will induce non-specific binding of cells to beads. DNase treatment of the cell suspension prior to cell isolation will prevent this problem without harming intact cells. Some sample tubes (e.g., glass or polystyrene) tend to bind cells non-specifically, which can be a major problem when working with minor cell populations (e.g., rare, circulating tumour cells). Precoating of sample tubes with a protein solution before use or the use of low-binding plastic tubes is recommended.
4. Phagocyte cells (e.g., monocytes) will bind and engulf beads if incubation is performed at temperatures above 2–8 °C.

7 Summary

This chapter has described the use of Dynabeads for cell isolation and expansion. In general, magnetic beads coated with specific antibodies can be used either for isolation or depletion of various cell types. Positive or negative cell isolation can be performed depending on the nature of the starting sample, the cell surface markers and the downstream application in question. Positive cell isolation is the method of choice for unprocessed samples such as whole blood, and for downstream molecular applications. Positive cell isolation can also be used for any downstream application after detachment and removal of the beads. Negative cell isolation is the method of choice when it is critical that cells of interest remain untouched, i.e., no antibodies have been bound to any cell surface markers on the cells of interest. Some cell populations can only be defined by multiple cell surface markers. Such populations of cells can be isolated by the combination of negative and positive cell isolation. By coupling Dynabeads with antibodies directed against cell surface activation molecules, the beads can be used both for isolation and expansion of the cells.

Dynabeads are currently used in two major clinical applications: 1) In the Isolex® 300i Magnetic Cell Selection System for CD34 Stem Cell Isolation – the only FDA-approved and CE-marked technology for the clinical selection of CD34+ cells, allowing the passive depletion of tumor cells from autologous apheresis products. 2) For ex vivo T cell isolation and expansion using Dynabeads *ClinExVivo* CD3/CD28 for clinical trials in novel adoptive immunotherapy.

References

1. Ugelstad J, Mørk PC, Herder Kaggerud K, Ellingsen T, Berge A (1980) Swelling of oligomer particles: New methods of preparation of emulsions and polymer dispersions. Adv Colloid Interface Sci 13:101
2. Ugelstad J, Kilaas L, Aune O, Bjørgum J, Herje R, Schmid R, Stenstad P, Berge A (1994) Monodisperse polymer particles. In: Uhlén M, Hornes E, Olsvik Ø (eds) Advances of Biomagnetic Separation. Eaton Publ Comp, Natick, MA, pp 1–20
3. Funderud S, Nustad K, Lea T, Vartdal F, Gaudernack G, Stenstad P, Ugelstad J (1987) Fractionation of lymphocytes by immunomagnetic beads. In: Klaus GGB (ed) Lymphocytes: A Practical Approach. IRL Press, Oxford, pp 55–65
4. Luxembourg AT, Borrow P, Teyton L, Brunmark AB, Peterson PA, Jackson MR (1998) Biomagnetic isolation of antigen-specific CD8+ T cells usable in immunotherapy. Nat Biotech 16:281–285
5. Marquez C, Trigueros C, Franco JM, Ramiro AR, Carrasco YR, Lopez-Botet M, Toribio ML (1998) Identification of a common developmental pathway for thymic natural killer cells and dendritic cells. Blood 91:2760–2771
6. Soltys J, Swain SD, Sipes KM, Nelson LK, Hanson AJ, Kantele JM, Jutila MA, Quinn MT (1999) Isolation of bovine neutrophils with biomagnetic beads: Comparison with standard Percoll density gradient isolation methods. J Immunol Meth 226:71–84
7. Chang CC, Ciubotariu R, Manavalan JS, Yuan J, Colovai AI, Piazza F, Lederman S, Colonna M, Cortesini R, Dalla-Favera R, Suciu-Foca N (2002) Tolerization of dendritic cells by T(S) cells: The crucial role of inhibitory receptors ILT3 and ILT4. Nat Immunol 3:237–243
8. Garlie NK, LeFever AV, Siebenlist RE, Levine BL, June CH, Lum LG (1999) T cells coactivated with immobilized anti-CD3 and anti-CD28 as potential immunotherapy for cancer. J Immunother 22:336-345
9. Lum LG, LeFever AV, Treisman JS, Garlie NK, Hanson JP Jr (2001) Immune modulation in cancer patients after adoptive transfer of anti-CD3/anti-CD28-costimulated T cells-phase I clinical trial. J Immunother 24(5):408–419
10. Liebowitz DN, Lee KP, June CH (1998) Costimulatory approaches to adoptive immunotherapy. Curr Opin Oncol 10(6):533–541
11. Levine BL, Cotte J, Small CC, Carroll RG, Riley JL, Bernstein WB, Van Epps DE, Hardwick RA, June CH (1998) Large-scale production of CD4+ T cells from HIV-1-infected donors after CD3/CD28 costimulation. J Hematother 7:437–448
12. Levine BL, Bernstein WB, Aronson NE, Schlienger K, Cotte J, Perfetto S, Humphries MJ, Ratto-Kim S, Birx DL, Steffens C, Landay A, Carroll RG, June CH (2002) Adoptive transfer of costimulated CD4+ T cells induces expansion of peripheral T cells and decreased CCR5 expression in HIV infection. Nat Med 8:47–53
13. Thomas AK, June CH (2001) The promise of T-lymphocyte immunotherapy for the treatment of malignant disease. Cancer J 7:S67–S75
14. Köhler G, Milstein C (1975) Continous cultures of fused cells secreting antibody of predefined specificity. Nature 256:495–497

15. Bosnes M, Bergholtz S, Borgnes A, Breivold E, Lycke K, Keiserud A, Borgen T, Lillehaug D (2002) Isolation of pure cells, Nuleic acids and Proteins from Single Cell Samples, Molecular and Cellur Proteomic 1–9 Addendum. HUPO First World Congress, Versailles France
16. Pisanic TR, Blackwell JD, Shubayev VI, Finones RR, Jin S (2007) Nanotoxicity of iron oxide nanoparticle nternalization in growing neurons. Biomaterials 28:2572–2581
17. Berry CC, Wells S, Charles S, Curtis ASG (2003) Dextran and albumin derivatised iron oxide nanoparticles: influence on fibroblasts in vitro. Biomaterials 24:4551–4557
18. Smeland EB, Funderud S, Kvalheim G, Gaudernack G, Rasmussen AM, Rusten L, Wang MY, Tindle RW, Blomhoff HK, Egeland T (1992) Isolation and characterization of human hematopoietic progenitor cells: an effective method for positive selection of CD34+ cells. Leukemia 6(8):845–852
19. Rasmussen AM, Smeland EB, Erikstein BK, Caignault L, Funderud S (1992) A new method for detachment of Dynabeads from positively selected B lymphocytes. J Immunol Meth 146(2):195–202
20. Baize S, Kaplon J, Faure C, Pannetier D, Georges-Courbot MC, Deubel V (2004) Lassa virus infection of human dendritic cells and macrophages is productive but failsto activate cells. J Immunol 172:2861–2869
21. Beckhove P, Feuerer M, Dolenc M, Schuetz F, Choi C, Sommerfeldt N, Schwendemann J, Ehlert K, Altevogt P, Bastert G, Schirrmacher V, Umansky V (2004) Specifically activated memory T cell subsets from cancer patients recognize and reject xenotransplanted autologous tumors. J Clin Invest 114:67–76
22. Friedl P, Noble PB, Zänker KS (1995) T lymphocyte locomotion in a three-dimensional collagen matrix: Expression and function of cell adhesion molecules. J Immunol 154:4973–4985
23. Chabannon C, Cornletta K, Lotz JP, Rosenfeld C, Shlomchik M, Yanovitch S, Marolleau JP, Sledge G, Srour EF, Burtness B, Camerlo J, Gravis G, Lee-Fischer J, Faucher C, Chabbert I, Krause D, Maraninchi D, Mills B, Kunkel L, Oldham F, Blaise D, Viens P (1998) High-dose chemotherapy followed by reinfusion of selected CD34+ peripheral blood cells in patients with poor-prognosis breast cancer: a randomized multicentre study. Brit J Cancer 7:913
24. Dreger P, Viehmann K, Von Neuhoff N, Glaubitz T, Petzold O, Glass B, Uharek L, Rautenberg P, Suttorp M, Mills B, Mitzky P, Schmitz B (1999) Autografting of highly purified peripheral blood progenitor cells following myeloablative therapy in patients with lymphoma: A prospective study of the long-term effects on tumor eradication, reconstitution of hematopoiesis, and immune recovery. Bone Marrow Transplant 24:153
25. Abonour R, Scott KM, Kunkel LA, Robertson MJ, Hromas R, Graves V, Lazaridis EN, Cripe L, Gharpure V, Traycoff CM, Mills B, Srour EF, Cornetta K (1998) Autologous transplantation of mobilized peripheral blood CD34+ cells selected by immunomagnetic procedures in patients with multiple myeloma. Bone Marrow Transplant 22:957
26. Rowley SD, Loken M, Radich J, Kunkel LA, Mills BJ, Gooley T, Holmberg L, Mcsweeney P, Beach K, Macleod B, Appelbaum F, Bensinger WI (1998) Isolation of CD34+ cells from blood stem cell components using the Baxter Isolex system. Bone Marrow Transplant 21:1253
27. Bensinger W, Appelbaum F, Rowley S, Storb R, Sanders J, Lilleby K, Gooley T, Demirer T, Schiffman K, Weaver C (1996) Factors that influence collection and engraftment of autologous peripheral blood stem cells. J Clin Oncol 13:2547

28. Falk M, Ussat S, Reiling N, Wesch D, Kabelitz D, Adam-Klages S (2004) Caspase inhibition blocks human T cell proliferation by suppressing appropriate regulation of IL-2, CD25 and cell cycle-associated proteins. J Immunol 173:5077–5085
29. Mandrekar P, Catalano D, Dolganiuc A, Kodys K, Szabo G (2004) Inhibition of myeloid dendritic accessory cell function and induction of T cell anergy by alcohol correlates with decreased IL-12 production. J Immunol 173:3398–3407
30. Sato K, Nakaoka T, Yamashita N, Yagita H, Kawasaki H, Morimoto C, Baba M, Matsuyama T (2005) TRAIL-transduced dendritic cells protect mice from acute graft-versus-host disease and leukemia relapse. J Immunol 174:4025–4033
31. Arruvito L, Sanz M, Banham AH, Fainboim L (2007) Expansion of CD4 primary-coated Dynabeads CD25= and FOXP3+ Regulatory T Cells during the Follicular Phase of the Menstrual Cycle: Implications for Human Reproduction. J Immunol 178:2572–2578
32. Jonuleit H, Schmitt E (2005) Regulatory T-cells in antitumor therapy: isolation and functional testing of CD4+CD25+ regulatory T-cells. Methods Mol Med 109:285–296
33. Hoffmann P, Eder R, Kunz-Schughart LA, Andreesen R, Edinger M (2004) Large-scale in vitro expansion of polyclonal human CD4+ CD25^{high} regulatory T cells. Blood 104(3):895–903
34. Levine BL, Mosca JD, Riley JL, Carroll RG, Vahey MT, Jagodzinski LL, Wagner KF, Mayers DL, Burke DS, Weislow OS, St Louis DC, June CH (1996) Antiviral effect and ex vivo CD4+ T cell proliferation in HIV-positive patients as a result of CD28 costimulation. Science 272:1939–1943
35. Levine BL, Bernstein WB, Connors M, Craighead N, Lindsten T, Thompson CB, June CH (1997) Effects of CD28 costimulation on long-term proliferation of CD4+ T cells in the absence of exogenous feeder cells. J Immunol 159:5921–5930
36. Carroll RG, Riley JL, Levine BL, Feng Y, Kaushal S, Ritchey DW, Bernstein W, Weislow OS, Brown CR, Berger EA, June CH, St Louis DC (1997) Differential regulation of HIV-1 fusion cofactor expression by CD28 costimulation of CD4+ T cells. Science 276:273–276
37. Levine BL, Cotte J, Small CC, Carroll RG, Riley JL, Bernstein WB, Van Epps DE, Hardwick RA, June CH (1998) Large-scale production of CD4+ T cells from HIV-1-infected donors after CD3/CD28 costimulation. J Hematother 7:437–448
38. Bonyhadi M, Frohlich M, Rasmussen A, Ferrand C, Grosmaire L, Robinet E, Leis J, Maziarz RT, Tiberghien P, Berenson RJ (2005) In vitro engagement of CD3 and CD28 corrects T cell defects in chronic lymphocytic leukemia. J Immunol 174:2366–2375
39. Parmar S, Robinson SN, Komanduri K, St John L, Decker W, Xing D, Yang H, McMannis J, Champlin R, de Lima M, Mollderm J, Rieber A, Bonyhadi M, Berenson R, Shpall EJ (2006) Ex vivo expanded umbilical cord blood T cells maintain naive phenotype and TCR diversity. Cytotherapy 8:149–157
40. Coito S, Sauce D, Duperrier A, Certoux JM, Bonyhadi ML, Collette A, Kuehlcke K, Herve P, Tiberghien P, Robinet E, Ferrand C (2004) Retrovirus-mediated gene transfer in human primary T lymphocytes induces an activation-and transduction/selection-dependent TCRBV repertoire skewing of gene-modified cells. Stem Cells Dev 13:71–81
41. Bonyhadi M, Frohlich M, Rasmussen A, Ferrand C, Grosmaire L, Robinet E, Leis J, Maziarz RT, Tiberghien P, Berenson RJ (2005) In vitro engagement of CD3 and CD28 corrects T cell defects in chronic lymphocytic leukemia. J Immunol 174:2366–2375
42. Hami L, Chana H, Craig S (2003) Comparison of a Static Process and a Bioreactor-based Process for the GMP Manufacture of Autologous Xcellerated T Cells for Clinical Trials. Bioprocess J 2(6):1–10

43. Casazza JP, Betts MR, Price DA, Precopio ML, Ruff LE, Brenchley JM, Hill BJ, Roederer M, Douek DC, Koup RA (2006) Acquisition of direct antiviral effector functions by CMV-specific CD4+ T lymphocytes with cellular maturation. J Exp Med 203:2865–2877
44. Kalamasz D, Long SA, Taniguchi R, Buckner JH, Berenson RJ, Bonyhadi M (2004) Optimization of Human T-Cell Expansion Ex Vivo Using Magnetic Beads Conjugated with Anti-CD3 and Anti-CD28 Antibodies. J Immunother 27:405–418
45. Dang Y, Knutson KL, Goodell V, dela Rosa C, Salazar LG, Higgins D, Childs J, Disis ML (2007) Tumor antigen-specific T-cell expansion is greatly facilitated by in vivo priming. Clin Cancer Res 13:1883–1891
46. Berger C, Blau AC, Clackson T, Riddell SR, Heimfeld S (2003) CD28 costimulation and immunoaffinity-based selection efficiently generate primary gene-modified T cells for adoptive immunotherapy. Blood 101:476–484
47. Ferrand C, Robinet E, Contassot E, Certoux J-M, Lim A, Herve P, Tiberghien P (2000) Retrovirus-Mediated Gene Transfer in Primary T Lymphocytes: Influence of the transduction/Selection Process and of ex Vivo Expansion on the T Cell Receptor b Chain Hypervariable Region Repertoire. Human Gene Therapy 11:1151–1164
48. Mitsuyasu RT, Anton PA, Deeks SG, Scadden DT, Connick E, Downs MT, Bakker A, Roberts MR, June CH, Jalali S, Lin AA, Pennathur-Das R, Hege KM (2000) Prolonged survival and tissue trafficking following adoptive transfer of CD4zeta gene-modified autologous CD4(+) and CD8(+) T cells in human immunodeficiency virus-infected subjects. Blood 96:785–793
49. Levine BL, Bernstein WB, Aronson NE, Schlienger K, Cotte J, Perfetto S, Humphries MJ, Ratto-Kim S, Birx DL, Steffens C, Landay A, Carroll RG, June CH (2002) Adoptive transfer of costimulated CD4+ T cells induces expansion of peripheral T cells and decreased CCR5 expression in HIV infection. Nat Med 8:47–53
50. Laport GG, Levine BL, Stadtmauer EA, Schuster SJ, Luger SM, Grupp S, Bunin N, Strobl FJ, Cotte J, Zheng Z, Gregson B, Rivers P, Vonderheide RH, Liebowitz DN, Porter DL, June CH (2003) Adoptive transfer of costimulated T cells induces lymphocytosis in patients with relapsed/refractory non-Hodgkin's lymphoma following CD34-selected hematopoietic cell transplantation. Blood 102:2004–2013
51. Rapoport AP, Levine BL, Badros A, Meisenberg B, Ruehle K, Nandi A, Rollins S, Natt S, Ratterree B, Westphal S, Mann D, June CH (2004) Molecular remission of CML after autotransplantation followed by adoptive transfer of costimulated autologous T cells. Bone Marrow Transplant 33:53–60
52. Godfrey WR, Ge YG, Spoden DJ, Levine BL, June CH, Blazar BR, Porter SB (2004) In Vitro Expanded Human CD4+CD25+ T Regulatory Cells can Markedly Inhibit Allogeneic Dendritic Cell Stimulated MLR Cultures. Blood 104:453–461
53. Godfrey WR, Spoden DJ, Ge YG, Baker SR, Liu B, Levine BL, June CH, Blazar BR, Porter SB (2005) Cord blood CD4+CD25+-derived T regulatory cell lines express FoxP3 protein and manifest potent suppressor function. Blood 105:750–758
54. Earle KE, Tang Q, Zhou X, Liu W, Zhu S, Bonyhadi ML, Bluestone JA (2005) In vitro expanded human CD4+CD25+ regulatory T cells suppress effector T cell proliferation. Clin Immunol 115:3–9
55. Karakhanova S, Munder M, Schneider M, Bonyhadi M, Ho AD, Goerner M (2006) Highly Efficient Expansion of Human CD4+CD25+ Regulatory T Cells for Cellular Immunotherapy in Patients with Graft-Versus-Host Disease. J Immunother 29:336–349
56. Bondanza A, Valtolina V, Magnani Z, Ponzoni M, Fleischhauer K, Bonyhadi M, Traversari C, Sanvito F, Toma S, Radrizzani M, La Seta-Catamancio S, Ciceri F, Bor-

dignon C, Bonini C (2006) Suicide gene therapy of graft-versus-host disease induced by central memory human T lymphocytes. Blood Mar 107(5):1828–1836
57. Noonan K, Matsui W, Serafini P, Carbley R, Tan G, Khalili J, Bonyhadi M, Levitsky H, Whartenby K, Borrello I (2005) Activated marrow-infiltrating lymphocytes effectively target plasma cells and their clonogenic precursors. Cancer Res 65:2026–2034
58. Porter DL, Levine BL, Bunin N, Stadtmauer EA, Luger SM, Goldstein S, Loren A, Phillips J, Nasta S, Perl A, Schuster S, Tsai D, Sohal A, Veloso E, Emerson S, June CH (2006) A phase 1 trial of donor lymphocyte infusions expanded and activated ex vivo via CD3/CD28 costimulation. Blood 107(4):1325–1331
59. Thompson JA, Figlin RA, Sifri-Steele C, Berenson RJ, Frohlich MW (2003) A phase I trial of CD3/CD28-activated T cells (Xcellerated T cells) and interleukin-2 in patients with metastatic renal cell carcinoma. Clin Cancer Res 9(10Pt1):3562–3570
60. Glode ML, Pantuck A, Higano CS, Meyer J, Hami L, Craig S, Berenson RJ, Frohlich MW (2004) A phase I/II trial of CD3/CD28 activated T cells (Xcellerated T Cells) in patients with hormone refractory prostate cancer. J Clin Oncol 22(Abstract):2549
61. Deeks SG, Wagner B, Anton PA, Mitsuyasu RT, Scadden DT, Huang C, Macken C, Richman DD, Christopherson C, June CH, Lazar R, Broad DF, Jalali S, Hege KM (2002) A Phase II Randomized Study of HIV-Specific T-Cell Gene Therapy in Subjects with Undetectable Plasma Viremia on Combination Antiretroviral Therapy. Mol Ther 5:788–797
62. Humeau LM, Binder GK, Lu X, Slepushkin V, Merling R, Echeagaray P, Pereira M, Slepushkina T, Barnett S, Dropulic LK, Carroll R, Levine BL, June CH, Dropulic B (2004) Efficient lentiviral vector-mediated control of HIV-1 replication in CD4 lymphocytes from diverse HIV+ infected patients grouped according to CD4 count and viral load. Mol Ther 9(6):902–913
63. Taylor PA, Panoskaltsis-Mortari A, Swedin JM, Lucas PJ, Gress RE, Levine BL, June CH, Serody JS, Blazar BR (2004) L-Selectin(hi) but not the L-selectin(lo) CD4+25+ T-regulatory cells are potent inhibitors of GVHD and BM graft rejection. Blood 104(12):3804–3812
64. Tang Q, Henriksen KJ, Bi M, Finger EB, Szot G, Ye J, Masteller EL, McDevitt H, Bonyhadi M, Bluestone JA (2004) In Vitro-expanded Antigen-specific Regulatory T Cells Suppress Autoimmune Diabetes. J Exp Med 199:1455–1465
65. Trenado A, Sudres M, Tang Q, Maury S, Charlotte F, Gregoire S, Bonyhadi M, Klatzmann D, Salomon BL, Cohen JL (2006) Ex vivo-expanded CD4+CD25+ immunoregulatory T cells prevent graft-versus-host-disease by inhibiting activation/differentiation of pathogenicT cells. J Immunol 176(2):1266–1273
66. Fowler D, Hou J, Foley J, Hakim F, Odom J, Castro K, Carter C, Read E, Gea-Banacloche J, Kasten-Sportes C, Kwak L, Wilson W, Levine B, June C, Gress R, Bishop M (2002) Phase I clinical trial of donor T-helper Type-2 cells after immunoablative, reduced intensity allogeneic PBSC transplant. Cytotherapy 4:429–430
67. Rapoport AP, Stadtmauer EA, Aqui N, Badros A, Cotte J, Chrisley L, Veloso E, Zheng Z, Westphal S, Mair R, Chi N, Ratterree B, Pochran MF, Natt S, Hinkle J, Sickles C, Sohal A, Ruehle K, Lynch C, Zhang L, Porter DL, Luger S, Guo C, Fang HB, Blackwelder W, Hankey K, Mann D, Edelman R, Frasch C, Levine BL, Cross A, June CH (2005) Restoration of immunity in lymphopenic individuals with cancer by vaccination and adoptive T-cell transfer. Nat Med 11(11):1162–1163
68. Kipps TJ, Castro JE, Wierda W, Keating MJ, Bole J, Meyer J, Roehrs H, Bouchard L, Yuan V, Chana H, Hami L, Bonyhadi M, Berenson RJ, Frohlich MW (2003) A phase I/II trial of Xcellerated T Cells in patients with Chronic Lymphocytic Leukemia (CLL). Blood 102(Abstract):370

69. Castro JE, Wierda WG, Kipps TJ, Keating MJ, Bole J, Anderson B, Meyer J, Bonyhadi M, Berenson RJ, Frohlich MW (2003) A phase I/II trial of Xcellerated T Cells in patients with Chronic Lymphocytic Leukemia. Blood 104(Abstract):2508
70. Colucci S, Brunetti G, Rizzi R, Zonno A, Mori G, Colaianni G, Del Prete D, Faccio R, Liso A, Capalbo S, Liso V, Zallone A, Grano M (2004) T cells support osteoclastogenesis in an in vitro model derived from human multiple myleloma bone disease: the role of the OPG/TRAIL interaction. Blood 104:3722–3730
71. Slager EH, van der Minne CE, Kruse M, Krueger DD, Griffioen M, Osanto S (2004) Identification of multiple HLA-DR-restricted epitopes of the tumor associated antigen CAMEL by CD4+ Th1/Th2 lymphocytes. J Immunol 172:5095–5102
72. Holm AM, Aukrust P, Aandahl EM, Muller F, Tasken K, Froland SS (2003) Impaired secretion of IL-10 by T cells from patients with common variable immunodeficiency-involvement of protein kinase A type I. J Immunol 170:5772-5777
73. Dardalhon V, Jaleco S, Rebouissou C, Ferrand C, Skander N, Swainson L, Tiberghien P, Spits H, Noraz N, Taylor N (2000) Highly efficient gene transfer in naive human T cells with a murine leukemia virus based vector. Blood 96:885-893
74. Aasheim HC, Delabie J, Finne EF (2005) Ephrin-A1 binding to CD4+ T lymphocytes stimulates migration and induces tyrosine phosphorylation of PYK2. Blood 105(7):2869–2876
75. Koh KP, Wang Y, Yi T, Shiao SL, Lorber MI, Sessa WC, Tellides G, Pober JS (2004) T cell-mediated vascular dysfunction of human allografts results from IFN-γ dysregulation of NO synthase. J Clin Investigation 114:846–856
76. Ranjbar S, Ly N, Thim S, Reynes JM, Goldfeld AE (2004) Mycobacterium tuberculosis recall antigens suppress HIV-1 replication in anergic donor cells via CD8+ T cell expansion and increased IL-10 levels. J Immunol 172(3):1953–1959
77. Zhang R, Fichtenbaum CJ, Hildeman DA, Lifso JD, Chougnet C (2004) CD40 Ligand Dysregulation in HIV Infection: HIV Glycoprotein 120 Inhibits Signaling Cascades Upstream of CD40 Ligand Transcription. J Immunol 172:2678–2686
78. Sharma P, Gnjatic S, Jungbluth AA, Williamson B, Herr H, Stockert E, Dalbagni G, Machele Donat S, Reuter VE, Santiago D, Chen YT, Bajorin DF, Ritter G, Old LJ (2003) Frequency of NY-ESO-1 and LAGE-1 expression in bladder cancer and evidence of a new NY-ESO-1 T-cell epitope in a patient with bladder cancer. Cancer Immun 3:19
79. Sloan D, Zahariadis G, Posavad CM, Pate NT, Kussick SJ, Jerome KR (2003) CTL are inactivated by herpes simplex virus-infected cells expressing a viral protein kinase. J Immunol 171:6733–6741
80. Qiuping Z, Qun L, Chunsong H, Xiaolian Z, Baojun H, Mingzhen Y, Chengming L (2003) Selectively Increased Expression and Functions of Chemokine Receptor CCR9 on CD4– T Cells from Patients with T-Cell Lineage Acute Lymphocytic Leukemia 1. Cancer Res 63:6469–6477
81. Lundi S, Johansson C, Svennerholm AM (2002) Oral Immunization with a *Salmonella enterica* Serovar Typhi Vaccine Induces Specific Circulating Mucosa-Homing CD4+ and CD8+ T Cells in Humans. Infect Immunity 70(10):5622–5627
82. Takedatsu H, Shichijo S, Katagiri K, Sawamizu H, Sata M, Itoh K (2004) Identification of peptide vaccine candidates sharing among HLA-A3+, -A11+, -A31+ and -A33+ cancer patients. Clin Cancer Res 10:1112–1120
83. Lundgren A, Suri-Payer E, Enarsson K, Svennerholm AM, Lundin BS (2002) Helicobacter pylori-specific CD4+CD25high regulatory T cells suppress memory T cell responses to *H. pylori* in infected individuals. Infect Immunity 71:1755–1762
84. Wagar EJ, Cromwell MA, Shultz LD, Woda BA, Sullivan JL, Hesselton RAM, Dale L, Greiner DL (2002) Regulation of human cell engraftment and development of EBV-

related lymphoproliferative disorders in Hu-PBL-scid mice. J Immunol 165:518–527
85. Piliero LM, Sanford AN, McDonald-McGinn DM, Zackai EH, Sullivan KE (2004) T-cell homeostasis in humans with thymic hypoplasia due to chromosome 22q11.2 deletion syndrome. Blood 103(3):1020–1025
86. Arruvito L, Sanz M, Banham AH, Fainboim L (2007) Expansion of CD4+CD25+and FOXP3+ Regulatory T Cells during the Follicular Phase of the Menstrual Cycle: Implications for Human Reproduction. J Immunol 178:2572–2578
87. Cosentino M, Fietta AM, Ferrari M, Rasini E, Bombelli R, Carcano E, Saporiti F, Meloni F, Marino F, Lecchini S (2007) Human CD4+CD25+ regulatory T cells selectively express tyrosine hydroxylase and contain endogenous catecholamines subserving an autocrine/paracrine inhibitory functional loop. Blood 109(2):632–642
88. Wing K, Larsson P, Sandström K, Lundin SB, Suri-Payer E, Rudin A (2005) CD4+CD25+FOXP3+ regulatory T cells from human thymus and cord blood suppress antigen-specific T cell responses. Immunology 115(4):516–525
89. Jiang S, Camara N, Lombardi G, Lechler RI (2003) Induction of allopeptide-specific human CD4+CD25+ regulatory T cells ex vivo. Blood 102:2180–2186
90. Satoguina JS, Weyand E, Larbi J, Hoerauf A (2005) T Regulatory-1 Cells Induce IgG4 Production by B Cells: Role of IL-10. J Immunol 174:4718–4726
91. Rice JS, Newman J, Wang C, Michael DJ, Diamond B (2005) Receptor editing in peripheral B cell tolerance. PNAS 102(5):1608–1613
92. Petlickovski A, Laurenti L, Li X, Marietti S, Chiusolo P, Sica S, Leone G, Efremov DG (2005) Sustained signaling through the B-cell receptor induces Mcl-1 and promotes survival of chronic lymphocytic leukemia B cells. Blood 105(12):4820–4827
93. Laurenti L, Petlickovski A, Rumi C, Gobessi S, Piccioni P, Tarnani M, Puggioni P, Marietti S, Sica S, Leone G, Efremov DG (2005) Comparison of ZAP-70/Syk mRNA levels with immunoglobulin heavy-chain gene mutation status and disease progression in chronic lymphocytic leukemia. Haematologica 90:1533–1540
94. Imadome K, Shirakata M, Shimizu N, Nonoyama S, Yamanashi Y (2003) CD40 ligand is a critical effector of Epstein–Barr virus in host cell survival and transformation. PNAS 100:7836–7840
95. Mulder A, Eijsink C, Kardol MJ, Franke-van Dijk ME, van der Burg SH, Kester M, Doxiadis II, Claas FH (2003) Identification, isolation and culture of HLAA2-specific B lymphocytes using MHC Class I tetramers. J Immunol 171:6599–6603
96. Rasmussen AM, Smeland E, Erikstein BK, Caignault L, Funderud S (1992) A new method for detachment of Dynabeads from positively selected B lymphocytes. J Immunol Meth 146:195–202
97. Grund EM, Spyropoulos DD, Watson DK, Muise-Helmericks RC (2005) IL-2 and IL-15 regulate ETS1 expression via ERK1/2 and MNK1 in human NK cells. J Biol Chem 280:4772–4778
98. Kai S, Goto S, Tahara K, Sasaki A, Tone S, Kitano S (2004) Indoleamine 2,3-dioxygenase is necessary for cytolytic activity of natural killer cells. Scand J Immunol 59:177–182
99. Igarashi T, Wynberg J, Srinivasan R, Becknell B, McCoy JP Jr, Takahashi Y, Suffredini DA, Linehan WM, Caligiuri MA, Childs RW (2004) Enhanced cytotoxicity of allogeneic NK cells with killer immunoglobulin-like receptor ligand incompatibility against melanoma and renal cell carcinoma cells. Blood 104:170–177
100. Kai S, Goto S, Tahara K, Sasaki A, Kawano K, Kitano S (2003) Inhibition of indoleamine 2,3-dioxygenase suppresses NK cell activity and accelerates tumor growth. J Exp Ther Oncol 3(6):336–345

101. Araki M, Kondo T, Gumperz JE, Brenner MB, Miyake S, Yamamura T (2003) Th2 bias of CD4+ NKT cells derived from multiple sclerosis in remission. Int Immunol 15(2):279–288
102. D'Ettore G, Forcina G, Andreotti M, Sarmati L, Palmisano L, Andreoni M, Vella S, Mastroianni CM, Vullo V (2004) Interleukin-15 production by monocyte-derived dendritic cells and T cell proliferation in HIV-infected patients with discordant response to highly active antiretroviral therapy. Clin Exp Immunol 135:280–285
103. Yu Q, Kovacs C, Yue FY, Ostrowski MA (2004) The role of the p38 mitogenactivated protein kinase, extracellular signalrelated kinase and phosphoinositide-3-OH kinase signal transduction pathways in CD40 ligand-induced dendritic cell activation and expansion of virus-specific CD8+ T cell memory responses. J Immunol 172:6047–6056
104. Li LQ, Liu D, Hutt-Fletcher L, Morgan A, Masucci MG, Levitsky V (2002) Epstein-Barr virus inhibits the development of dendritic cells by promoting apoptosis of their monocyte precursors in the presence of granulocyte macrophage-colony-stimulating factor and interleukin-4. Blood 99:3725–3734
105. Jahnsen FL, Strickland DH, Thomas JA, Tobagus IT, Napoli S, Zosky GR, Turner DJ, Sly PD, Stumbles PA, Holt PG (2006) Accelerated Antigen Sampling and Transport by Airway Mucosal Dendritic Cells following Inhalation of a Bacterial Stimulus. J Immunol 177:5861–5867
106. Palucka AK, Laupeze B, Aspord C, Saito H, Jego G, Fay J, Paczesny S, Pascual V, Banchereau J (2005) Immunotherapy via dendritic cells. Adv Exp Med Biol 560:105–114
107. Shortman K, Liu YJ (2002) Mouse and human dendritic cell subtypes. Nat Rev Immunol 2(3):151–161
108. Weissman D, Houping N, Scales D, Dude A, Capodici J, McGibney K, Abdool A, Isaacs SN, Cannon G, Kariko K (2000) HIV Gag mRNA Transfection of Dendritic Cells (DC) Delivers Encoded Antigen to MHC Class I and II Molecules, Causes DC Maturation, and Induces a Potent Human In Vitro Primary Immune Response. J Immunol 165:4710–4717
109. Chia-Chun J, Wright A, Punnonen J (2000) Monocyte-Derived CD1a+ and CD1a-Dendritic Cell Subsets Differ in Their Cytokine Production Profiles, Susceptibilities to Transfection, and Capacities to Direct Th Cell Differentiation. J Immunol 165:3584–3591
110. Moharita AL, Taborga M, Corcoran KE, Bryan M, Patel PS, Rameshwar P (2006) SDF-1alpha regulation in breast cancer cells contacting bone marrow stroma is critical for normal hematopoiesis. Blood 108(10):3245–3252
111. Canque B, Camus S, Dalloul A, Kahn E, Yagello M, Dezutter-Dambuyant C, Schmitt D, Schmitt C, Gluckman JC (2000) Characterization of dendritic cell differentiation pathways from cord blood CD34(+)CD7(+)CD45RA(+) hematopoietic progenitor cells. Blood 96(12):3748–3756
112. Ueda T, Yoshino H, Kobayashi K, Kawahata M, Ebihara Y, Ito M, Asano S, Nakahata T, Tsuji K (2000) Hematopoietic repopulating ability of cord blood CD34(+) cells in NOD/Shi-scid mice. Stem Cells 18(3):204–213
113. Egeland T, Gaudernack G (1994) CD34 The Gateway to the Study of Lymphohematopoietic Progenitor and Leukemic Cells. The Immunologist 2:65–70
114. Tong J, Hoffman R, Siena S, Srour EF, Bregini M, Gianni AM (1994) Characterization and quantitation of primitive hematopoietic progenitor cells present in peripheral blood autografts. Exp Hematol 22(10):1016–1024

115. Sutherland DR, Keating A (1992) The CD34 Antigen: Structure, Biology and Potential Clinical Applications. J Haematother 1:115–129
116. Cao TM, Shizuru JA, Wong RM, Sheehan K, Laport GG, Stockerl-Goldstein KE, Johnston LJ, Stuart MJ, Grumet FC, Negrin RS, Lowsky R (2005) Engraftment and survival following reducedintensity allogeneic peripheral blood hematopoietic cell transplantation is affected by CD8+ T cell dose. Blood 105:2300–2306
117. Beckhove P, Feuerer M, Dolenc M, Schuetz F, Choi C, Sommerfeldt N, Schwendemann J, Ehlert K, Altevogt P, Bastert G, Schirrmacher V, Umansky V (2004) Specifically activated memory T cell subsets from cancer patients recognize and reject xenotransplanted autologous tumors. J Clin Invest 114:67–76
118. Thickett DR, Armstrong L, Millar AB (2002) A role for vascular endothelial growth factor in acute and resolving lung injury. Am J Respir Crit Care Med 166(10):1332–1337
119. Murray J, Ward C, O'Flaherty JT, Dransfield I, Haslett C, Chilvers ER, Rossi AG (2003) Role of leukotrienes in the regulation of human granulocyte behaviour: dissociation between agonist-induced activation and retardation of apoptosis. Br J Pharmacol 139(2):388–398
120. Blom M, Tool AT, Mul FP, Knol EF, Roos D, Verhoeven AJ (1995) Eosinophils isolated with two different methods show different characteristics of activation. J Immunol Meth 178:183–193
121. Bach MK, Brashler JR, Sanders ME (1990) Preparation of large numbers of highly purified normodense human eosinophils from leukapheresis. J Immunol Meth 130:277–281
122. Gibbs BF, Noll T, Falcone FH, Haas H, Vollmer E, Vollrath I, Wolff HH, Amon U (1997) A three-step procedure for the purification of human basophils from buffy coat blood. Inflamm Res 46(4):137–142
123. de Belder A, Radomski M, Hancock V, Brown A, Moncada S, Martin J (1995) Megakaryocytes from patients with coronary atherosclerosis express the inducible nitric oxide synthase. Arterioscler Thromb Vasc Biol 15(5):637–641
124. Pasquet JM, Dachary-Prigent J, Nurden AT (1998) Microvesicle release is associated with extensive protein tyrosine dephosphorylation in platelets stimulated by A23187 or a mixture of thrombin and collagen. Biochem J 333(Pt3):591–599
125. Sanmugalingam D, Wardlaw AJ, Bradding P (2000) Adhesion of human lung mast cells to bronchial epithelium: evidence for a novel carbohydrate-mediated mechanism. J Leukoc Biol 68(1):38–46
126. Hewett PW, Murray JC (1993) Immunomagnetic purification of human microvessel endothelial cells using Dynabeads coated with monoclonal antibodies to PECAM-1. Eur J Cell Biol 62:451–454
127. Hewett PW, Murray JC (1993) Human microvessel endothelial cells: isolation, culture and characterization. In Vitro Cell Dev Biol Anim 29A(11):823–830
128. Hewett PW, Murray JC, Price EA, Watts ME, Woodcock M (1993) Isolation and characterization of microvessel endothelial cells from human mammary adipose tissue. In Vitro Cell Dev Biol Anim 29A(4):325–331
129. Lalor PF, Edwards S, McNab G, Salmi M, Jalkanen S, Adams DH (2002) Vascular Adhesion Protein-1 Mediates Adhesion and Transmigration of Lymphocytes on Human Hepatic Endothelial Cells. J Immunol 169:983–992
130. Rigolin GM, Fraulini C, Ciccone M, Mauro E, Bugli AM, De Angeli C, Negrini M, Cuneo A, Castoldi G (2006) Neoplastic circulating endothelial cells in multiple myeloma with 13q14 deletion. Blood 107(6):2531–2535

131. Woywodt A, Blann AD, Kirsch T, Erdbrugger U, Banzet N, Haubitz M, Dignat-George F (2006) Isolation and enumeration of circulating endothelial cells by immunomagnetic isolation: proposal of a definition and a consensus protocol. Thromb Haemost 4(3):671–677
132. Woywodt A, Goldberg C, Scheer J, Regelsberger H, Haller H, Haubitz M (2004). An improved assay for enumeration of circulating endothelial cells. Ann Hematol 83(8):491–494
133. Koumas L, King AE, Critchley HO, Kelly RW, Phipps RP (2001) Fibroblast heterogeneity: existence of functionally distinct Thy 1(+) and Thy 1(-) human female reproductive tract fibroblasts. Am J Pathol 159(3):925–935
134. Fernandez-Shaw S, Shorter SC, Naish CE, Barlow DH, Starkey PM (1992) Isolation and purification of human endometrial stromal and glandular cells using immunomagnetic microspheres. Hum Reprod 7(2):156–161
135. Werther K, Normark M, Hansen BF, Brunner N, Nielsen HJ (2000) The use of the CELLection kit in the isolation of carcinoma cells form mononuclear cell suspensions. J Immunol Meth 238(1–2):133–141
136. Kielhorn E, Schofield K, Rimm DL (2002) Use of magnetic enrichment for detection of carcinoma cells in fluid specimens. Cancer 94(1):205–211
137. Hardingham JE, Kotasek D, Sage RE, Eaton MC, Pascoe VH, Dobrovic A (1995) Detection of circulating tumour cells in colorectal cancer by immunobead-PCR is a sensitive prognostic marker for relapse of disease. Molec Med 1(7):789–794
138. de Cremoux P, Extra JM, Denis MG, Pierga JY, Bourstyn E, Nos C, Clough KB, Boudou E, Martin EC, Muller A, Pouillart P, Magdelenat H (2000) Detection of MUC1-expressing mammary carcinoma cells in the peripheral blood of breast cancer patient by real-time polymerase chain reaction. Clin Cancer Res 6(8):3117–3122
139. Okabe M, Matzno S, Nagira M, Ying X, Kohama Y, Mimura T (1992) Collection of acrosome-reacted human sperm using monoclonal antibody-coated paramagnetic beads. Mol Reprod Dev 32(4):389–393
140. Pederson L, Kremer M, Judd J, Pascoe D, Spelsberg TC, Riggs BL, Oursler MJ (1999) Androgens regulate bone resorption activity of isolated osteoclasts in vitro. Proc Natl Acad Sci 96(2):505–510
141. Aukrust P, Muller F, Svardal AM, Ueland T, Berge RK, Froland SS (2003) Disturbed glutathione metabolism and decreased antioxidant levels in human immunodeficiency virus-infected patients during highly active antiretroviral therapy–potential immunomodulatory effects of antioxidants. J Infect Dis 188(2):232–238
142. Xu Z, Butfiloski EJ, Sobel ES, Morel L (2004) Mechanisms of peritoneal B-1A cells accumulation induced by murine lupus susceptibility locus Sle2. J Immunol 173:6050–6058
143. Metwali A, Blum AM, Elliott DE, Setiawan T, Weinstock JV (2004) Cutting Edge: Hemokinin has Substance P-like function and expression in inflammation. J Immunol 172:6528–6532
144. Kamath AT, Sheasby CE, Tough DF (2005) Dendritic Cells and NK Cells Stimulate Bystander T Cell Activation in Response to TLR Agonists through Secretion of IFN-$\alpha\beta$ and IFN-γ. J Immunol 174:767–776
145. Glennie S, Soeiro I, Dyson PJ, Lam EW, Dazzi F (2005) Bone marrow mesenchymal stem cells induce division arrest anergy of activated T cells. Blood 105(7):2821–2827
146. Chandra AP, Ouyang L, Yi S, Wong JK, Ha H, Walters SN, Patel AT, Stokes R, Jardine M, Hawthorne WJ, O'Connell PJ (2007) Chemokine and toll-like receptor signaling in macrophage mediated islet xenograft rejection. Xenotransplantation 14(1):48–59

147. Henderson WR Jr, Chi EY, Bollinger JG, Tien YT, Ye X, Castelli L, Rubtsov YP, Singer AG, Chiang GK, Nevalainen T, Rudensky AY, Gelb MH (2007) Importance of group X-secreted phospholipase A2 in allergen-induced airway inflammation and remodeling in a mouse asthma model. J Exp Med 204(4):865–877
148. Harrington LE, Hatton RD, Mangan PR, Turner H, Murphy TL, Murphy KM, Weaver CT (2005) Interleukin 17-producing CD4+ effector T cells develop via a lineage distinct from the T helper type 1 and 2 lineages. Nat Immunol 6(11):1123–1132
149. Anderson G, Jenkinson EJ, Moore NC, Owen JJ (1993) MHC class II-positive epithelium and mesenchyme cells are both required for T-cell development in the thymus. Nature 362:70–73
150. von Garnier C, Filgueira L, Wikstrom M, Smith M, Thomas JA, Strickland DH, Holt PG, Stumbles PA (2005) Anatomical Location Determines the Distribution and Function of Dendritic Cells and Other APCs in the Respiratory Tract. J Immunol 175:1609–1618
151. Puliaev R, Nguyen P, Finkelman FD, Via CS (2004) Differential requirements for IFN-g in CTL maturation in acute murine graft-versus host disease. J Immunol 173:910–919
152. Gorbachev AV, Fairchild RL (2004) CD40 engagement enhances antigen-presenting Langerhans cell priming of IFN-gproducing CD4+ and CD8+ T cells independently of Il-12. J Immunol 173(4):2443–2452
153. Rus V, Nguyen V, Puliaev R, Puliaeva I, Zernetkina V, Luzina I, Papadimitriou JC, Via CS (2007) T Cell TRAIL Promotes Murine Lupus by Sustaining Effector CD4 Th Cell Numbers and by Inhibiting CD8 CTL Activity. J Immunol 178:3962–3972
154. Roan NR, Gierahn TM, Higgins DE, Starnbach MN (2006) Monitoring the T cell response to genital tract infection. PNAS 103(32):12069–12074
155. Kutzler MA, Robinson TM, Chattergoon MA, Choo DK, Choo AY, Choe PY, Ramanathan MP, Parkinson R, Kudchodkar S, Tamura Y, Sidhu M, Roopchand V, Kim JJ, Pavlakis GN, Felber BK, Waldmann TA, Boyer JD, Weiner DB (2005) Coimmunization with an Optimized IL-15 Plasmid Results in Enhanced Function and Longevity of CD8 T Cells That Are Partially Independent of CD4 T Cell Help. J Immunol 175(1):112–123
156. Stephens LA, Gray D, Anderton SM (2005) CD4+CD25+ regulatory T cells limit the risk of autoimmune disease arising from T cell receptor crossreactivity. PNAS 102(48):17418–17423
157. Giorda E, Sibilio L, Martayan A, Feriotto G, Bianchi N, Mischiati C, Di Rosa F, Pozzi L, Gambari R, Giacomini P (2005) Modular usage of the HLA-DRA promoter in extra-hematopoietic and hematopoietic cell types of transgenic mice. FEBS J 272:3214–3226
158. Zeytin HE, Patel AC, Rogers CJ, Canter D, Hursting SD, Schlom J, Greiner JW (2004) Combination of a Poxvirus-Based Vaccine with a Cyclooxygenase-2 Inhibitor (Celecoxib) Elicits anti-tumor Immunity and Long-Term Survival in CEA.Tg/MIN Mice. Cancer Res 64:3668–3678
159. O'Leary JG, Goodarzi M, Drayton DL, von Andrian UH (2006) T cell-and B cell-independent adaptive immunity mediated by natural killer cells. Nat Immunol 7(5):507–516
160. Vremec D, Pooley J, Hochrein H, Wu L, Shortman K (2000) CD4 and CD8 Expression by Dendritic Cell Subtypes in Mouse Thymus and Spleen. J Immunol 164:2978–2986
161. Martinez del Hoyo G, Lopez-Bravo M, Metharom P, Ardavin C, Aucouturier P (2006) Prion Protein Expression by Mouse Dendritic Cells Is Restricted to the Nonplasmacytoid Subsets and Correlates with the Maturation State. J Immunol 177:6137–6142

162. Rui T, Cepinskas G, Feng Q, Ho YS, Kvietys PR (2001) Cardiac myocytes exposed to anoxia-reoxygenation promote neutrophil transendothelial migration. Am J Physiol Heart Circ Physiol 281(1):H440–447
163. Balconi G, Spagnuolo R, Dejana E (2000) Development of endothelial cell lines from embryonic stem cells: A tool for studying genetically manipulated endothelial cells in vitro. Arterioscler Thromb Vasc Biol 20(6):1443-1451
164. Camerer E, Kataoka H, Kahn M, Lease K, Coughlin SR (2002) Genetic evidence that protease-activated receptors mediate factor Xa signaling in endothelial cells. J Biol Chem 277(18):16081-1608
165. Tiruppathi C, Freichel M, Vogel SM, Paria BC, Mehta D, Flockerzi V, Malik AB (2002) Impairment of store-operated Ca2+ entry in TRPC4(-/-) mice interferes with increase in lung microvascular permeability. Circ Res 91(1):70–76
166. Galvez BG, Genis L, Matias-Roman S, Oblander SA, Tryggvason K, Apte SS, Arroyo AG (2005) Membrane type 1-matrix metalloproteinase is regulated by chemokines monocyte-chemoattractant protein-1/ccl2 and interleukin-8/CXCL8 in endothelial cells during angiogenesis. J Biol Chem 280(2):1292–1298
167. Ross R, Ross XL, Ghadially H, Lahr T, Schwing J, Knop J, Reske-Kunz AB (1999) Mouse langerhans cells differentially express an activated T cell-attracting CC chemokine. J Invest Dermatol 113(6):99–998
168. Nishida S, Hosen N, Shirakata T, Kanato K, Yanagihara M, Nakatsuka S, Hoshida Y, Nakazawa T, Harada Y, Tatsumi N, Tsuboi A, Kawakami M, Oka Y, Oji Y, Aozasa K, Kawase I, Sugiyama H (2006) AML1-ETO rapidly induces acute myeloblastic leukemia in cooperation with the Wilms tumor gene, WT1. Blood 107(8):3303–3312
169. Raslova H, Baccini V, Loussaief L, Comba B, Larghero J, Debili N, Vainchenker W (2006) Mammalian target of rapamycin (mTOR) regulates both proliferation of megakaryocyte progenitors and late stages of megakaryocyte differentiation. Blood 107(6):2303–2310
170. Ghinassi B, Sanchez M, Martelli F, Amabile G, Vannucchi AM, Migliaccio G, Orkin SH, Migliaccio AR (2007) The hypomorphic Gata1low mutation alters the proliferation/differentiation potential of the common megakaryocytic-erythroid progenitor. Blood 109(4):1460–1471

Adv Biochem Engin/Biotechnol (2007) 106: 75–99
DOI 10.1007/10_2006_042

Published online: 16 December 2006

Affinity Adsorption of Cells to Surfaces and Strategies for Cell Detachment

John Hubble

School of Chemical Engineering, University of Bath, Claverton Down, Bath BA2 7AY, UK
j.hubble@bath.ac.uk

Abstract The use of bio-specific interactions for the separation and recovery of biomolecules is now widely established and in many cases the technique has successfully crossed the divide between bench and process scale operation. Although the major specificity advantage of affinity-based separations also applies to systems intended for cell fractionation, developments in this area have been slower. Many of the problems encountered result from attempts to take techniques developed for molecular systems and, with only minor modification to the conditions used, apply them for the separation of cells. This approach tends to ignore or at least trivialise the problems, which arise from the heterogeneous nature of a cell suspension and the multivalent nature of the cell/surface interaction.

To develop viable separation processes on a larger scale, effective contacting strategies are required in separators that also allow detachment or recovery protocols that overcome the enhanced binding strength generated by multivalent interactions. The effects of interaction valency on interaction strength needs to be assessed and approaches developed to allow effective detachment and recovery of adsorbed cells without compromising cell viability. This article considers the influence of operating conditions on cell attachment and the extent to which multivalent interactions determine the strength of cell binding and subsequent detachment.

Keywords Attachment · Bio-specific · Elution · Fractionation · Multivalent · Shear

Abbreviations

b	Half channel height (m)
c	Cell concentration (m^{-3})
D_c	Cell diffusivity ($m^2\,s^{-1}$)
f	Frictional coefficient ($kg\,s^{-1}\,mol^{-1}$)
f_b	Force per bond (mN)
F	Shape factor
F_t	Total applied force (mN)
h	H-hs (m)
hs	Separation distance (m)
H	Maximum separation distance (m)
j	Cell deposition rate ($m^{-2}\,s^{-1}$)
k_b	Boltzman constant ($J\,molecule^{-1}\,K^{-1}$)
k_f	Forward rate constant (s^{-1})
k_r	Reverse rate constant (s^{-1})
k_r'	Apparent reverse rate constant (s^{-1})
K_a	Association constant
K_d	Dissociation constant (M)
lc	Ligand concentration ($mol\,m^{-2}$)
nb	Number of bonds between cell and surface
nb_{max}	Average maximum possible number of bonds between cell and surface
nb_{min}	Minimum number of bonds to resist an applied force
N	Bond turn-over (s^{-1})
Pe	Peclet number
Q	Volumetric flow ($m^3\,s^{-1}$)
r_a	Radius of contact area (m)
r_p	Cell radius (m)
r^*	Cell radius to edge of contact area (m)
t	Time interval (s)
T	Temperature (K)
u	Linear velocity ($m\,s^{-1}$)
w	Channel width (m)
x	Distance from channel entrance (m)
α	Contact angle with surface
θ	Contact angle with surface
μ	Viscosity (Pa s)
γ	Bond interaction distance (m)
γ_L	Liquid–air interfacial tension ($n\,m^{-1}$)
τ_w	Wall shear stress (Pa)

1 Introduction

The basic concept of any bio-specific or affinity-based separation is to use the interaction of a biological receptor with its complementary ligand for the adsorption of a molecule or cell to an adsorbent surface. Conditions are chosen to take advantage of the selectivity of these interactions and their low dissociation constant such that immobilised ligands can be used to effectively adsorb the desired component from dilute solutions. Once bound, the adsorbed material can be washed to remove unbound species prior to inducing dissociation such that the purified product can be selectively recovered in a more highly concentrated form [1].

While selectivity or specificity is important, the magnitude of the dissociation constant for the interaction is also critical. The dissociation constant (K_d) is defined in terms of the ratio of unbound to bound species at equilibrium, and it follows that the amount bound is a simple function of the dissociation constant and adsorbent capacity such that the lower the dissociation constant the higher the fraction of product that can be bound during the loading stage.

The significance of K_d in determining product losses during the washing stage also needs to be considered for molecular separations. As washing proceeds, the liquid phase concentration of product will be reduced along with the contaminants to be removed, thus the product concentration in the liquid phase is dragged towards zero. This means that the equilibrium between bound and free product will be displaced and product will desorb. The extent of desorption will be a function of the wash volume used and the dissociation constant for the interaction. Essentially, the lower the dissociation constant the larger the wash volume that can be employed for a given level of product loss. Hence higher purification factors should be possible with interactions showing a lower dissociation constant [2].

When it comes to examination of the elution stage the requirements for the ideal dissociation constant change. At this point the binding equilibrium must be displaced to allow the release of adsorbed product. If the original dissociation constant is too low this will be difficult to achieve without the use of conditions that might damage the product.

Taken together, these factors mean that the ideal dissociation constant for a given molecular separation will be a compromise between the requirements for the loading and washing stages and those required for effective elution. These general considerations would be expected to apply to any affinity recovery system. However, when multivalent interactions can occur it is no longer possible to quantify adsorption in terms of a simple intrinsic dissociation constant.

When considering the interaction of multivalent particles with surfaces the definitions of concentration are significantly different to those used to describe molecular adsorption. The concentration of cell particles expressed in term of cell number per unit volume is analogous to molecular concentration and, together with the flow regime used, will determine the deposition rate of cells on a surface. However, the occurrence and strength of any interaction formed, once contact has occurred, will be dependent on the surface density and binding properties of ligands and receptors on the cell and adsorption surface. For an effective cell separation it is necessary for the cell surface interaction to retain the target cells while unwanted contaminants are removed in a washing stage. Once contaminants have been removed, adsorbed cells need to be detached using conditions that do not excessively compromise viability.

In most molecular affinity separation systems there are two main approaches to inducing elution. The most elegant approach is to use competing free ligand to displace the binding equilibrium, i.e. bio-specific elution. Alternatively, non-specific elution can be used where process conditions are changed such that the bio-specific interaction is disrupted. Typically, salt concentration, pH or temperature is used for this purpose.

While most molecular interactions used in affinity-based separations are monovalent, the adsorption of cells or particles to surfaces involves multiple interactions. Geometric considerations can be used together with ligand density and the size of the interacting species to determine the number of specific binding interactions or bonds (*nb*) that can occur between cells and surfaces [3]. Given a typical ligand and receptor density range of 10^6–10^8 binding sites m^{-2} and a contact area between 10^{-14}–10^{-12} m^2 this suggests that *nb* could have values between 1 and 10 000. For effective cell detachment it is necessary for the majority of these bonds to be simultaneously dissociated. This will not occur simply as a result of a dynamic equilibrium shift, and unless the binding conditions are significantly changed force will be required to effect recovery [4], greatly complicating the elution and recovery of affinity adsorbed cells.

There have been a number of detailed mechanistic studies of the effect of applied force on bonds holding cells to surfaces during natural cell adhesion processes. This has lead to a detailed modelling approach termed "adhesive dynamics" proposed by Hammer's group to describe cell adhesion under flow conditions in terms of the behaviour of some types of blood cells in the vicinity of a surface [5–8]. In the adhesive dynamics model, the rate expressions used are based on the treatment of adhesion molecules as springs. Depending on the relationship between rate of bond breaking and spring extension three classes of bonds are proposed: slip bonds (spring extension increases rate of breakage), catch bonds (spring extension decreases rate of breakage) and ideal bonds (rate of breakage is independent of spring extension). The specific ability of a receptor to form different bond types is used as a rationale for the fact that these cells roll along the adsorbing surface as an intermediate stage before full attachment. This is most commonly observed with interactions mediated by receptors of the selectin family whose properties are suggested to be specifically adapted to this role [7]. While it has been reported that antibody–antigen interactions between cells and a surface can also support rolling adhesion, this occurs only over a limited range of receptor density and range of shear stress [9], and it is likely that the complex relationship between extension and the rate of bond breakage seen with natural cell adhesion molecules may reflect their biological role rather than a more general property of biological interactions.

Once cells have reached stationary attachment, subsequent detachment is complicated by the multivalent nature of the interaction as described below. In practice, the difficulty of recovering affinity adsorbed cells in separation processes has led to the development and rapid expansion of techniques based on immuno-magnetic separations. In this case bio-specific interactions are used to couple magnetic particles to the target cell population, which can subsequently be recovered using a magnetic field. This technique is both simple and effective to use; however, the costs make it more attractive for diagnostic applications [10], although some comparative evaluations have been conducted with a view to scale up [11]. Overcoming the problem of cell recovery is probably the biggest challenge to be overcome if effective cell separation strategies are to be developed for use on a larger scale [12].

2 Theoretical Considerations

The simplest approach to modelling the interactions between cell and surface is to treat cells as solid spheres with ligand or receptors dispersed over their surface and to consider their binding to a planar surface supporting a distributed population of complementary ligands or receptors. This assessment of the attachment process results in a number of interdependent stages involving the interaction of fluid dynamic, mechanical force and biochemical effects [13]. For interaction to occur cells must first be brought into contact with a surface. The resultant "deposition rate" will be a function of the fluid flow conditions over the surface and the particle size and geometry [14].

Once in contact with the surface, the cell is able to form interactions dependent on the surface characteristics of both cell and attachment surface. Taking the simplified case of a ligand/receptor-mediated attachment, where secondary non-specific interactions cannot occur, the number of bonds that can form will be a function of ligand and receptor densities. If each bond requires a specific force to break it, the number of bonds between cell and surface will determine the shear stress that the attached cell will be able to resist [15]. The importance of this effect is shown in the case of neutrophil recruitment to the wall of a damaged blood vessel, where attachment is stimulated by the release of a soluble messenger that causes the neutrophil to express higher surface concentrations of the appropriate receptor. This increases number of affinity interactions possible between the cells and the receiving surface and increases the applied force that an attached cell is able to resist. In this case, complete attachment is preceded by conditions where cells transiently attach and detach in the rolling phenomenon. This slows the cells progressively to the point where they finally make a firm attachment [16].

Quantitative assessment of the shear stress values compatible with attachment and those required to detach pre-adsorbed cells shows that there is an order of magnitude difference [17]. Work in our laboratory has shown that the shear stress required to generate detachment increases with incubation time up to a maximum value, suggesting that additional interactions are progressively formed after the initial cell attachment [18].

Depending on the application envisaged, different aspects of the attachment/detachment process might assume greater or lesser significance. For example, investigation of seeding strategies for use in tissue culture applications will be concerned predominantly with contact and adhesion. These stages are also important in determining the progression of many infectious diseases where pathogen attachment to a target cell is critical and has led to studies aimed at developing agents that can inhibit the attachment process [19, 20]. Work on the significance of cell rolling in inflammation usually includes assessment of both attachment and detachment as essential steps in the rolling process [5]. Similarly effective cell separation processes will require careful assessment of both attachment and detachment phenomena.

Hubble [21] described an integrated model based on a process engineering approach to account for both the effects of fluid flow and changes in the density and binding parameters of the ligand/receptor interaction. The model used standard flow correlations to determining both the frequency of cell surface collisions and the consequential effects of surface shear in determining the behaviour of cells after contact. These include surface rolling and time-dependent binding stabilisation.

It is assumed that differences in attachment resulting from the binding interactions between ligand and receptor stem solely from differences in the intrinsic rate constants that describe bond association and dissociation coupled with the effect of applied force on the dissociation constant. No additional assumptions were made with respect to bond properties.

2.1
Cell Deposition

Correlations between fluid dynamic conditions and cell deposition rates have been reported for a number of experimental systems. These include biofilm development [22], the interaction of charged polystyrene particles with quartz surfaces [14] and membrane separations [23].

A number of these reports have used a deposition equation based on the Smoluchowski–Levich approximation [14]. For a parallel plate channel this can be written as:

$$j = \frac{D_c c}{0.89 r_p} \left(\frac{2}{9} \frac{bPe}{x} \right)^{\frac{1}{3}} , \tag{1}$$

where j is the cell deposition rate, r_p the cell radius, D_c the cell diffusivity, c the cell concentration, b the half channel height, and x the distance from channel entrance.

The Peclet number (Pe) for a parallel plate geometry flow chamber is given by:

$$Pe = \frac{3Qr_p^3}{4wb^3D_c}, \tag{2}$$

where Q is the volumetric flow rate and w the channel width.

Correlations for determination of cell diffusivity are based on the Stokes relationship [23]:

$$D_c = \frac{k_b T}{f}, \tag{3}$$

where k_b is the Boltzman constant and T is temperature.

The frictional coefficient f can be determined from Stokes law and the Perrin shape factor:

$$f = 6\pi\mu r_p F, \tag{4}$$

where F is the shape factor and μ the viscosity.

2.2 Stabilisation of Attachment

Once in contact with a surface, it is possible for cell attachment to either be stabilised by the formation of additional bonds or destabilised by the loss of existing bonds. Hubble [24] proposed a steady state model for cell/surface interactions that used a dimensionless association constant to describe the number of bonds formed between cell and surface once an initial interaction had formed. The rationale used was that the initial attachment spatially constrains the positional relationship between cell and surface such that volume-averaged concentration terms are inappropriate (Fig. 1). This concept has been extended to describe the progressive detachment of cells from surfaces and is capable of predicting the steady state retained fraction of cells as a function of applied shear [25]. While these reports are limited to consideration of steady state, the approach can be extended to describe the kinetics of bond formation by considering the balance of forward (k_f) and reverse (k_r) rate constants:

$$nb \underset{k_r}{\overset{k_f}{\rightleftharpoons}} nb + 1, \tag{5}$$

where nb is the number of bonds.

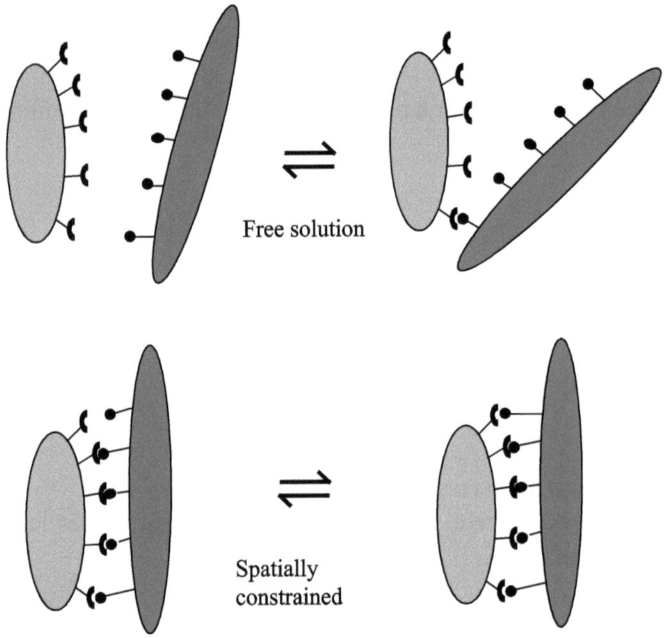

Fig. 1 Schematic showing the effect of the initial interaction to spatially constrain subsequent interactions in multivalent binding (from [12])

This is analogous to the assumption made by Chang and Hammer [6] that the binding reaction is a first order process. In their case the assumption made is that the apparent first order forward rate constant is given by the product of a second order intrinsic forward rate constant and the number of free receptors. Here the assumption is that spatial constraints dictate that each receptor will only be able to interact with a single ligand. Hence, there will be a finite population of potential ligand receptor pairs within the cell surface contact area, and these associate and dissociate in a first order process.

The force acting on the cell is usually assumed to have no effect on the association rate. The effect of applied force on the bond dissociation rate is quantified using the relationship proposed by Bell [4] to relate the apparent reverse constant to the intrinsic rate constant:

$$k'_{\mathrm{r}} \exp\left(\frac{\gamma F_{\mathrm{t}}}{k_{\mathrm{b}} T n b}\right), \tag{6}$$

where γ is the bond interaction distance.

As the number of interacting groups is small, description of the time course of bond formation was predicted for individual cell-surface interactions using a stochastic approach based on probability theory [26]. Given a probability (P) of there being j bonds at time t, the aim is to determine the probability that at a time $t + \Delta t$ the number of bonds has reached a new value n. Using this approach the stochastic change in the number cell/surface interactions for a given cell can be predicted as a function of time.

2.3 Strength of Cell Attachment

To quantify the force needed to detach an affinity adsorbed cell it is necessary to determine the number of bonds formed with the surface and the strength of each individual bond. The maximum number of interactions between a cell and surface will be a function of the surface density of ligand and receptor, the radius of the attaching cell and the maximum separation distance. Geometrical considerations, assuming a spherical cell, allow the formulation of an expression to describe the radius of the contact area (r_a) between cell and surface (Fig. 2) [15]:

$$r_a = r_p \sqrt{1 - \frac{(r_p - H)^2}{r_p^2}}, \tag{7}$$

where H is the maximum separation distance.

The contact area and the lower of either the surface receptor or ligand concentration determine the theoretical maximum number of bonds.

If each bond generates a specific attachment force, the number of bonds can be used to estimate the force required to detach a cell from the surface. This will also depend on the size of the cell and the radius of the contact area

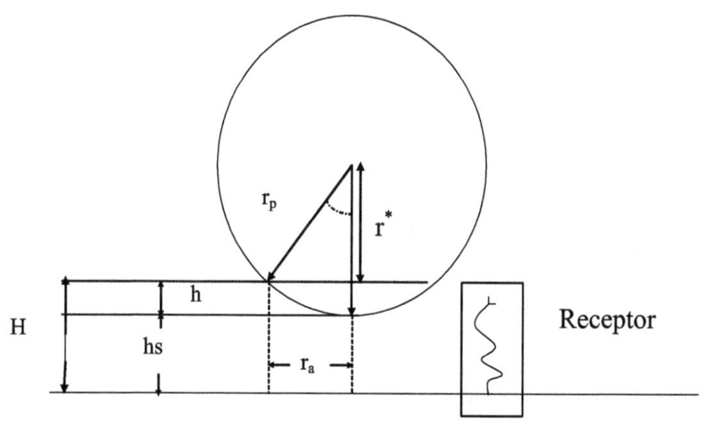

Fig. 2 Geometrical relationship between cell radius and receptor bond length in determining the contact area for a cell–surface interaction. The maximum number of interactions is a function of the surface concentration of ligand and receptor, the radius of the attaching particle and the maximum separation distance (bond length) (from [15])

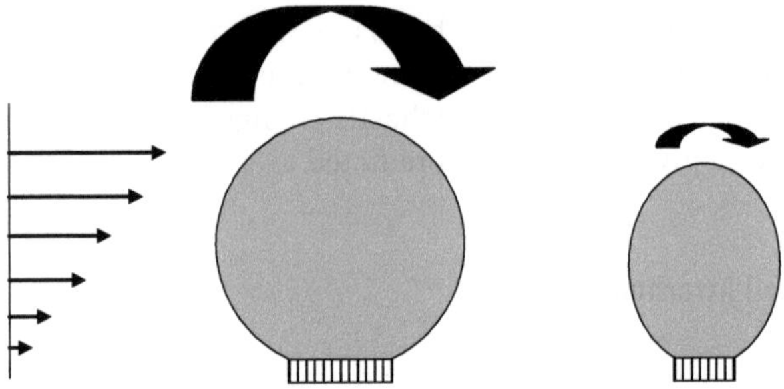

Fig. 3 Schematic showing the effects of size and contact area on the ability of a cell to resist an applied force

(Fig. 3). For the parallel flow chamber geometry commonly used to study cell surface interactions, Cozens-Roberts et al. [15] concluded that the total force (F_t) applied to a cell by fluid flow could be reduced to the following simple expression:

$$F_t = \frac{110 r_p^3 \tau_w}{r_a} . \tag{8}$$

In a parallel plate channel the wall shear stress is given by:

$$\tau_w = \frac{6u\mu}{2b} , \tag{9}$$

where b is the half channel height, u the linear velocity and μ the viscosity.

Using these relationships the force required to detach an affinity adsorbed cell from a surface can be determined and hence the flow conditions required to effect detachment can be quantified [21, 27].

While the equations developed above consider the effects of multiple interactions on binding it is likely that kinetic effects during the loading stage might result in a steady state bond number dependent on the conditions used. Even if the theoretical maximum number of interactions is determined by geometrical and binding considerations there is no guarantee that this maximum will be achieved (Fig. 4). Unless the ligand and receptor distribution are both uniform over the contact area it is probable that the number of bonds that form will be dependent on the orientation of the initial contact and the rate at which interactions subsequently occur. The situation might be regarded as analogous to annealing, where the ultimate crystal structure of a metal is dependant on the rate at which it is cooled from a molten state. Certainly there is a large volume of data that show multivalent interactions are time-stabilised [28–30]. In many cases this results from cellular responses [31]. However, results in our laboratory obtained with both artificial cell mimics and yeasts show that there can also be a time-dependent

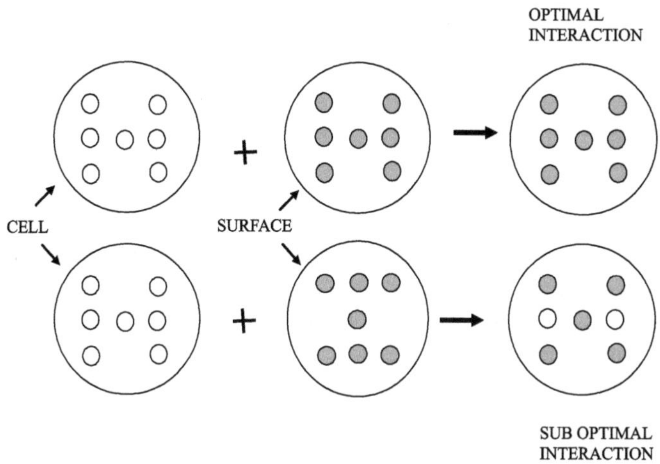

Fig. 4 Ligand/receptor-mediated interaction of cell with surface showing the importance of orientation in determining number of interacting ligand/receptor pairs. (N.B. even in the optimal orientation it cannot be assumed that the theoretical number of interactions will be reached) (Redrawn from [12])

stabilisation effect in systems where such re-orientation of binding molecules is not possible [18].

2.4 Model Formulation

The surface site density used for modelling studies was calculated from the surface concentrations of receptor and ligand. Simulations were conducted using surface concentrations in the range quoted for bead studies ($10^{14}-10^{16}\ \mathrm{m}^{-2}$) by Cozens-Roberts et al. [15]. The value chosen sets the maximum value for the number of interactions (nb_{max}) within the contact area calculated from Eq. 7.

Once the bond number has been calculated it is compared with the minimum number of bonds (nb_{min}) needed to match the applied force as calculated from Eq. 8:

$$nb_{\mathrm{min}} = \frac{F_{\mathrm{t}}}{f_{\mathrm{b}}} \,. \qquad (10)$$

Only if the number of bonds formed is greater than nb_{min} does the cell attach. The integrated model treats a surface element as a square grid of attachment sites to track the progressive attachment and detachment of cells. This grid is mapped to an area equal to ten times that, which results from the maximum number of adsorbed cells (i.e maximum fractional coverage of 0.1). The cell deposition rate to this area is determined using Eqs. 1–4. Once the position and number of bonds holding each of the attached cells have been determined for the attachment period, the number of bonds attaching each cell is revised using the stochastic numerical integration. At the start of the next time interval, the number of bonds holding each cell is compared with nb_{min}. Thus the change in the number of bonds in each time interval can be predicted as a function of both rate constants for the interaction and applied shear stress (Fig. 5). If the number of bonds has fallen below the minimum needed to appose the applied force the cell is deemed to have detached. This sequence is then repeated for the next time interval. The model tracks both number of attached cells, and the average number of bonds forming the attachment, as a function of time.

Using this modelling approach it is possible to predict the effect of changes in binding properties, cell dimensions, ligand and receptor densities and flow conditions on the attachment, stabilisation and detachment phases of a cell separation cycle. For example, Fig. 6 shows the effect of increasing volumetric flow rate through a parallel plate adsorber. As flow is increased the cell attachment rate increases with flow, reflecting an increased deposition rate. However, as flow is increased further, surface shear effects increase the competing detachment rate and net attachment falls.

Figure 7 shows that for conditions where net bond formation is positive the average number of bonds between cells and surface continues to increase

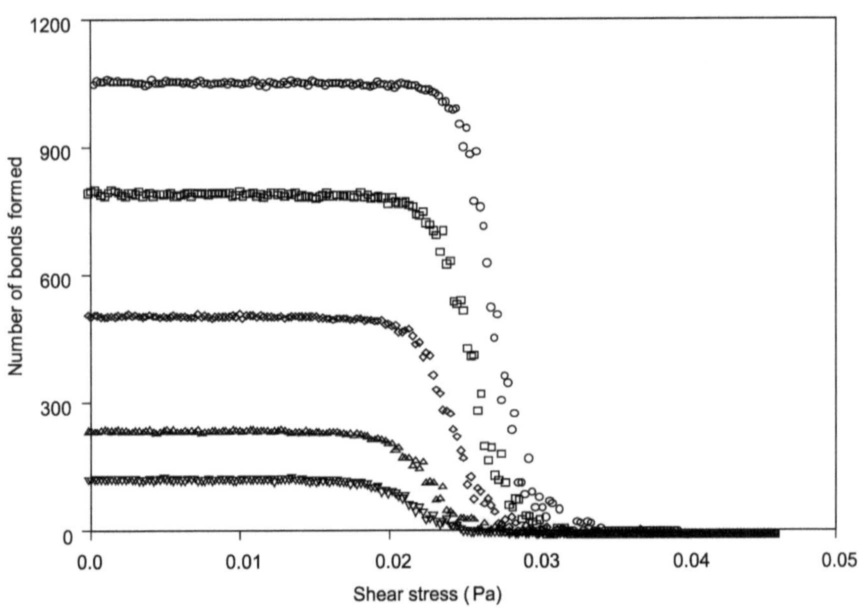

Fig. 5 Effect of shear stress (Pa) on number of bonds formed in 1 s. Initial contact is set to one bond. Order of magnitude of rate constant k_r ($k_f = 1.15k_r$ in each case): ○ 1, □ 0.5, ◇ 0.25, △ 0.1, ▽ 0.05 (from [21])

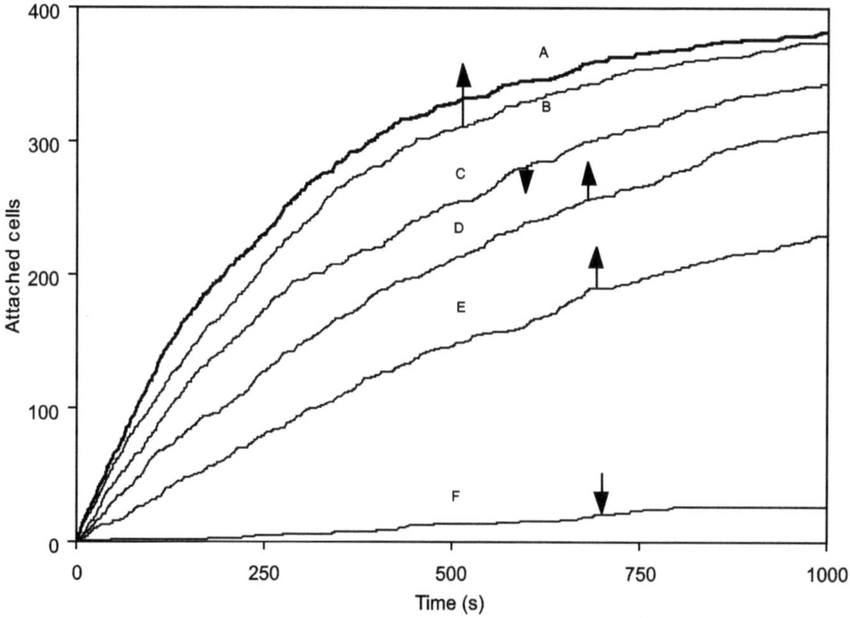

Fig. 6 Effect of volumetric flow rate ($m^3 s^{-1}$) on a simulated cell attachment time course in a thin channel adsorber. *A* 1.25×10^{-8}, *B* 8.3×10^{-9}, *C* 1.67×10^{-8}, *D* 8.3×10^{-10}, *E* 1.67×10^{-10}, *F* 2.1×10^{-8}. *Arrows* indicate whether trend is increasing or decreasing with flow rate (from [21])

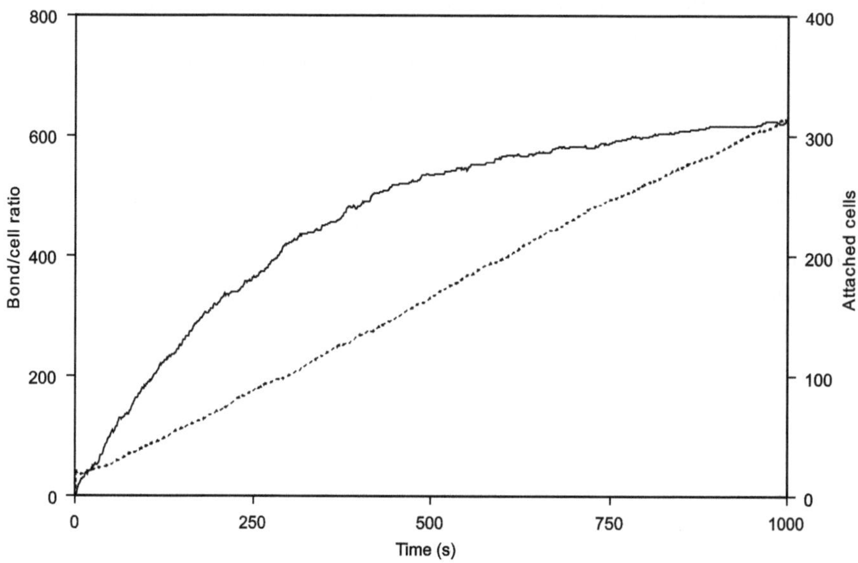

Fig. 7 Simulated time course for cell attachment (*solid line*) and bond/cell ratio (*broken line*) (from [21])

once the attached cell fraction is stable, reflecting the experimentally observed time-dependent stabilisation.

3
Cell Detachment Protocols

Given that the final number of bonds between cell and surface may be determined by sequence as well as by the extent of reaction, the attachment strength can be controlled by the loading conditions as well as the ligand density used. For example, it is possible that the loading stage can be moderated by the use of temperature or the presence of competitive monovalent inhibitors to give the ultimate binding strength required. Indeed a very close analogy to annealing could be obtained by physically contacting the cells in a high concentration of inhibitor, which is then progressively reduced as binding proceeds. Therefore, the first step in developing an effective detachment strategy is to control the attachment stage to ensure that the attachment force is limited to the minimum needed for effective capture and washing.

In molecular affinity chromatography systems, elution will be induced by some change to the chemical environment of the adsorbent. This could be the use of a bio-specific competitor where competing free ligand is used to displace the binding equilibrium. The problems here are that, as discussed below, multivalent competitors may be required, the cost of the eluting agent may be prohibitive, and the rate of desorption is often slow as it will be dictated by the "off" constant of the interaction.

To overcome these problems in molecular separations the process conditions are commonly changed such that the interaction is disrupted. The requirement is that while the changes induce a sufficiently large structural change to promote desorption, they are not sufficiently harsh that they cause denaturation or product damage. The feasibility of this approach depends on the strength of the binding interaction and if K_d is too low excessively harsh conditions may be required. This problem is particularly severe for separations intended for use with mammalian cells where the reagents used may be cytotoxic or lead to excessive changes in osmotic pressure.

An alternative chemical approach to detachment, which can be used to recover both molecules and cells in affinity separation protocols, is to use a proteolytic digestion to cleave either the ligand or receptor. In the case of genetically engineered proteins this can be facilitated by the inclusion of a "fusion tag" containing an amino acid sequence tailored to the specificity of a particular protease [32]. For cell detachment, digestion of the ligand or receptor is possible [33]. A potential disadvantage of this approach is that the adsorbent cannot be reused.

3.1
Bio-specific Elution

In the case of bio-specific elution, the concentration of a monovalent competitor would need to be determined with respect to the apparent dissociation constant of the multivalent interaction. In systems where nb can be large this would lead to concentrations that are not physically achievable. The situation can be improved by the use of multivalent competitive inhibitors [34].

Although multivalent competitors have not been reported for use in affinity separations, studies conducted on cell–cell interactions have shown that multivalent inhibitors can be many times as effective as an equivalent monovalent species [35, 36]. However, although these findings have significant therapeutic implications the cost of such preparations would make them unattractive for use in cell separation systems. In addition, as with molecular separations, the kinetics of bio-specific displacement is likely to be slow.

3.2
Non-specific Elution

The requirement here is that while the conditions induce a sufficiently large effect to promote desorption, they are not harsh enough to cause denaturation. The feasibility of meeting this constraint is dependent on the dissociation constant for the specific interaction.

Although the cooperative effects of multivalent interactions significantly reduce the apparent dissociation constant for the overall interaction, it is the individual interaction constant that determines the chemical conditions needed to non-specifically disrupt binding. This suggests that non-specific eluents that increase the ligand/receptor dissociation constant will be more effective than bio-specific competitors. The constraint on their use is the extent to which cell viability is compromised by elevated reagent concentrations or significant pH shifts; this suggests that engineering solutions are required to allow effective delivery strategies that minimise toxic side effects.

At the bulk concentrations needed to effect desorption, chemical toxicity may compromise the use of non-specific eluents. However, if cells are adsorbed onto a porous surface such as an ultrafiltration membrane, non-specific eluents can be diffused or convectively passed through the membrane while a tangential flow is maintained above the membrane. This allows the formation of an eluent concentration gradient from the membrane surface down to the bulk concentration. Balancing of transmembrane and tangential flows allows desorption to be controlled in terms of applied force, localised eluent concentration in the attachment region and extent of bulk eluent dilution (Fig. 8). The principle of this approach has been success-

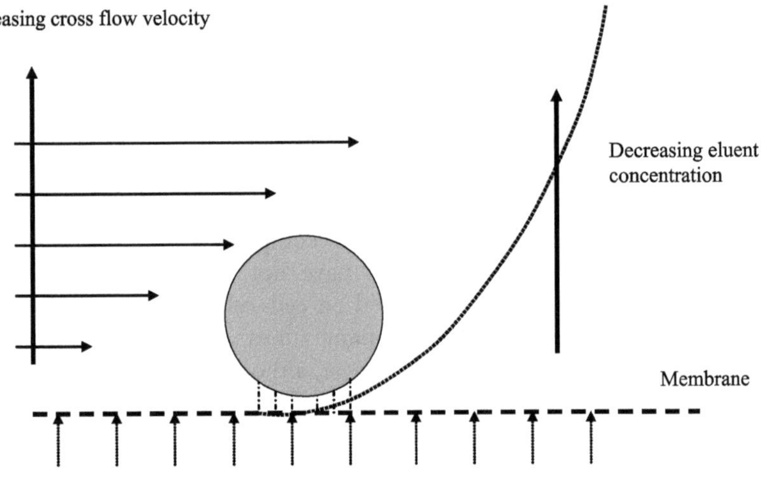

Fig. 8 Effect of cross-flow velocity on the concentration gradient of a non-specific eluent delivered through a porous adsorption surface

fully demonstrated by Mandrusov et al. [37] using diffusion of hydrochloric acid through a dialysis membrane acting as the ligand support to effect elution.

3.3 Shear-Induced Detachment

The effects of force on an attached cell have been described above and the relationship between force and fluid flow described for thin channel geometry. Providing the detachment forces required do not lead to cell damage this approach has the advantage of requiring no chemical changes to the suspension media. However, care has to be taken to ensure that the flow regime does not lead to excessive dilution of the recovered cell population.

In cell-based systems an external force in excess of the attachment force holding the cell to the surface can be applied to generate physical detachment [38]. In practice, some applied force will be necessary to remove the cell from the surface, even if all the bonds were to be dissociated (in this situation cell recovery would be analogous to simple re-suspension). The amount of force required above this minimum will depend on the cell surface interactions. The constraints on using flow-induced forces are the level of cell damage and the degree of eluted cell dilution that result.

The use of gas bubbles to displace adherent particles from surfaces has been investigated in a number of cleaning studies aimed at electronics applications [39–41] and more recently in studies of bacterial adhesion [42]. These studies showed that surface tension-induced detachment forces arising from a passing liquid–air interface are several orders of magnitude larger than the hydrodynamic forces for colloidal particles with radii below 5 μm [39].

The maximal detachment force $F_{\max}$ generated by the passage of an air–liquid interface on an attached particle is given as [40]:

$$F_{\max} = 2\pi r_{\mathrm{p}} \gamma_{\mathrm{L}} \sin^2\left(\frac{\alpha}{2}\right) \cos\theta \,, \tag{11}$$

for $\theta < 90°$.

$$F_{\max} = -\, 2\pi r_{\mathrm{p}} \gamma_{\mathrm{L}} \sin^2\left(\frac{\pi + \alpha}{2}\right) \cos\theta \,, \tag{12}$$

for $\theta > 90°$,
where r_{p} is the particle radius, γ_{L} is the liquid–air interfacial tension, and θ and α are contact angles on the collector and the particle surface, respectively.

The major advantage of gas as a displacing agent is that high shear stresses can be generated without the need for high liquid throughput, minimising cell dilution. The choice of an appropriate gas also minimises extraneous reagent contamination to the system.

4
Considerations for Effective Separator Design

There are two central factors that emerge from the assessment of the fundamentals of cell/surface interactions, namely the significance of multivalency, and the importance of applied force. This suggests that protocols based on the use of packed beds where shear forces are constrained, and bio-specific elution where elevated eluent concentrations are needed, are unlikely to be successful.

Taking the physical design of the contactor first, the fundamental considerations dictate a geometry that allows effective contacting of cells with surface and the subsequent potential to control the force applied to attached cells by varying the fluid shear at the surface.

It is immediately apparent that a fixed bed adsorber configuration does not well satisfy these requirements. In addition to the problems of steric entrapment identified by Sharma and Mahendroo [43], the compressible nature of most adsorbent beads coupled with the pressure drop/flow relationships inherent in fixed bed operation will place unacceptable limits on the displacement force that can be applied to attached cells. However, recent developments in the area of molecular adsorption include the use of expanded beds for the processing of particulate-containing crude feeds [44]. In this case, many of the problems of steric entrapment will be alleviated by the more open structure of the expanded bed. Chase's group at Cambridge have reported on the use of expanded bed protocols for a number of cell separations [45,46]. In their work yeast cells were adsorbed on a concanavalin A derivatised perfluorocarbon support. The adsorbent was used to give a settled height of 20 cm and fluidised at $0.7\,\mathrm{cm^3\,min^{-1}}$ to give an expanded height of 40 cm. *Saccharomyces cerevisiae* suspension was applied at a concentration of 1.8×10^8 cells $\mathrm{cm^{-3}}$ until 20% breakthrough. The bed was then washed with buffer and eluted in expanded mode.

Adsorption kinetics were attractive, showing a time constant of ≤ 8 min; however, while adsorbed cells could be eluted using 500 mM methyl α,D-mannopyranoside as a competitor, the kinetics of release were slowed by the multipoint nature of the interaction.

An alternative approach, which offers a number of attractive possibilities, is the use of a membrane-based system. In addition to separation based on the primary sieving properties of membranes, a number of workers have investigated their use as a combined sieve/adsorber [47]. In many molecular separations the potential benefits of this combined approach are compromised by fouling effects, which can lead to rapid changes in apparent molecular weight cut-off. However, the membrane adsorber is particularly well suited for use in cell separations where the potential to adjust the balance between trans-membrane flux and fluid velocity parallel to the surface means that adsorption, wash and elution steps can be carried out under separate, optimised conditions (Fig. 9).

In practice, a membrane-based approach offers the closest analogy to the flow cell geometries typically used in cell/surface detachment studies, such that flow conditions can easily be designed to match those identified in assay systems. In addition to this advantage, the ability to generate a convective flow normal to the membrane surface allows effective control of the contacting/adsorption stage of the separation (Fig. 9a). For cell separations the membrane pore size can be such that concentration polarisation and fouling by macromolecules can be minimised, and while biocompatibility issues may be important, the issue of mammalian cell stability has been addressed in the context of membrane bioreactor design [48].

The suitability of membrane separators for the wash and elution stages of separation also stem from the ease with which surface shear can be varied and again there is a large amount of background information, in this case resulting from the use of cross flow-induced shear to minimise membrane concentration polarisation and fouling [23]. There are two additional factors that increase the attractiveness of membrane separators. The first is the problem of non-specific detachment of target cells during the washing stage. Even though it may be possible to wash with shear rates below that needed to detach specifically bound cells they can still be removed as a result of collisions with contaminant cells washed from the surface. The ability to use a membrane with either a continuous or periodic back flux offers the potential to move cells away from the surface after detachment such that the potential for collision-induced detachment is reduced (Fig. 9b). However, probably the most significant advantage offered by membrane-based cell separators is the potential they offer for the effective use of non-specific eluents. In traditional fixed bed configurations the concentration of non-specific eluents required are likely to be highly cytotoxic and, in the case of mammalian cells, cause unacceptably large changes in osmotic pressure. However, as shown by Mandrusov et al. [37], a membrane system allows precise control on eluant delivery such that adjusting tangential and trans-membrane flows allows elution (Fig. 9c). In this study a cellophane membrane was mounted between two flow channels allowing the flow of buffered saline over the surface containing pre-adsorbed cells while hydrochloric acid (pH 1) was passed over the other side in a counter-current direction. Using this protocol the

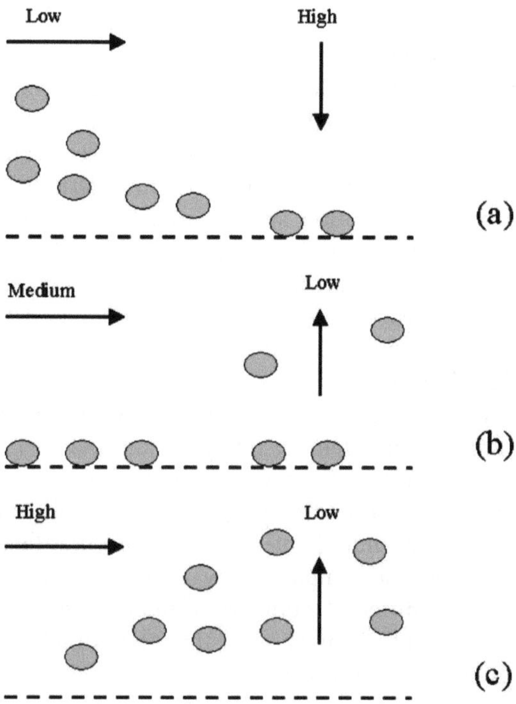

Fig. 9 Schematic of a membrane affinity surface used for cell separation. **a** Adsorption/ loading: Low cross-flow velocity to minimise shear induced detachment – high transmembrane velocity to deposit cells on surface, i.e. high concentration polarisation. **b** Wash: Medium cross flow to generate sufficient surface shear to detach cells held solely by non-specific interactions – low transmembrane velocity in the reverse direction (backflush) to minimise collision effects. **c** Elute: High cross flow to generate sufficient shear to detach affinity adsorbed cells. Low backflush to allow introduction of non-specific eluent (from [12])

authors found that all cells were removed within 15 min and that the viability of eluted cells was 60%. Direct elution using saline at pH 2 and 5 passed over the cell supporting surface resulted in lower viabilities of 10 and 30%, respectively.

Alternatives to chemical-based elution protocols can be based on the use of mechanical force to elute attached cells. This can be generated using fluid flow as described by workers studying cell attachment strength in parallel plate flow chamber systems, e.g. [5, 6, 15, 34]. However, for systems intended for cell fractionation, the use of fluid flow will either lead to excessive dilution or would require the recirculation of eluent through a pump, thereby exposing detached cells to elevated shear forces. Ujam et al. [46] employed "a crude attempt at an elution" of monocytes bound to a particulate immuno affinity adsorbent by removing the monocyte-loaded particles from an expanded bed column as slurry and then agitating in a beaker with a magnetic follower at 400 rpm for 10 min. While this resulted in recovery of monocytes at a yield and purity of 77 and 90%, respectively, the viability was reduced to 65% as a result of membrane damage.

An alternative approach employed in our laboratory was to use bubble-induced shear forces, as described above, to detach cells adsorbed to the lumen of nylon tubes [49]. While parallel plate cells are convenient for studying cell attachment phenomena, tubular surface configurations offer significant advantage in terms of adsorption surface area to volume ratios. Furthermore, the use of small diameter tubes greatly facilitates the introduction of bubbles having precise and reproducible dimensions. The tube cell used is described in Fig. 10. In each case, the separator was loaded by filling concanavalin A coated tubes with a suspension of yeast cells and then incubating to allow adsorption to occur.

The detachment resulting from the introduction of air bubbles into the tube bundle showed a consistent recovery of just below 1×10^8 cells over four loading cycles for a 0.5 mL bubble, with slightly higher recoveries for 1 and 2 mL bubbles. As each bubble passes over the cells, they are exposed to the passage of two air–liquid interfaces. As the interfaces generate the detachment shear stress, a comparison was made between different flow velocities (Fig. 11) and different numbers of bubbles for the same total injected air volume (Fig. 12). This showed that the cumulative effect of $4 \times$ 0.5 mL bubbles is consistently greater than a single 2 mL bubble. However, little additional benefit was gained after the passage of two bubbles. As with most elution protocols involving adsorptive separation, a trade-off must be made between fractional recovery and final concentration. The results obtained suggest that bubble detachment efficiency varies between 40 and 100% depending on bubble size and frequency. A follow-up study using affinity adsorbed erythrocytes showed essentially similar results (Figs. 13 and 14), with little evidence of physical damage evident during microscopic examination [50].

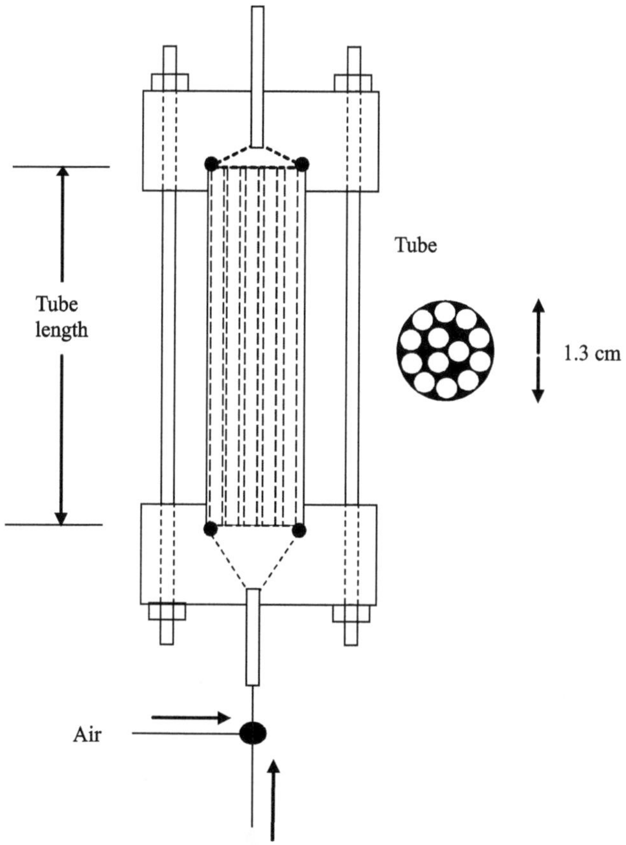

Fig. 10 Construction of a tubular cell adsorption chamber for use with bubble detachment of affinity adsorbed cells (from [49])

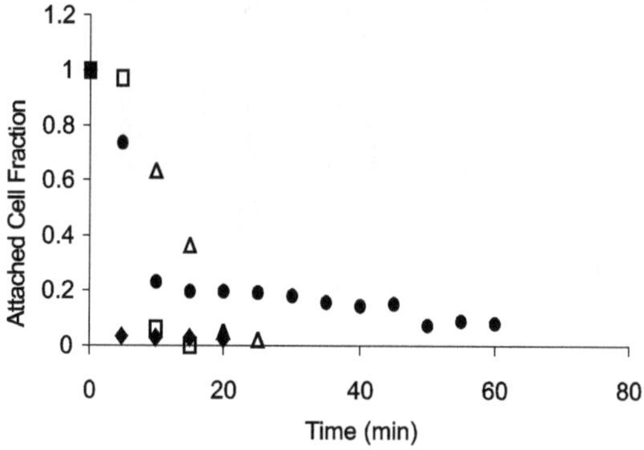

Fig. 11 Effect of bubble volume on cell detachment curves. Detachments were carried out with pH 6 phosphate buffer at flow velocity of 0.79 cm s^{-1} (shear stress 0.677 dyne cm^{-2}). Air volumes injected at 2-min intervals (from [49]): ♦ 2 mL, □ 1 mL, △ 0.5 mL, • 0.25 mL

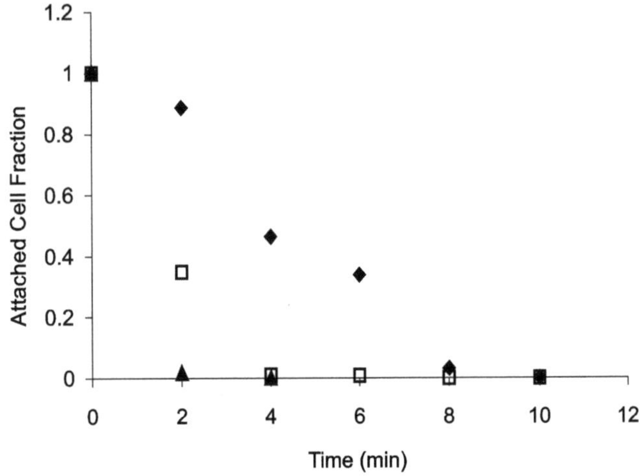

Fig. 12 Effect of flow velocity on cell detachment carried out with pH 6 phosphate buffer with 1 mL air injected into the flow system every 2 min (from [49]): ◆ 0.57 cm s^{-1} (shear stress 0.48 dyne cm^{-2}), □ 0.95 cm s^{-1} (shear stress 0.81 dyne cm^{-2}), ▲ 1.9 cm s^{-1} (shear stress 1.63 dyne cm^{-2})

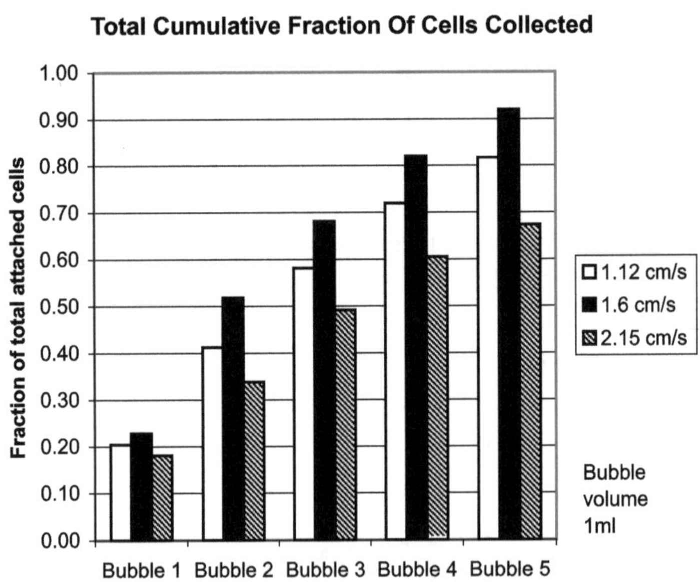

Fig. 13 Total cumulative fraction affinity adsorbed erythrocytes detached in a tubular adsorber as a function of number of bubbles passed over the surface at three different flow velocities (constant bubble size) (from [50])

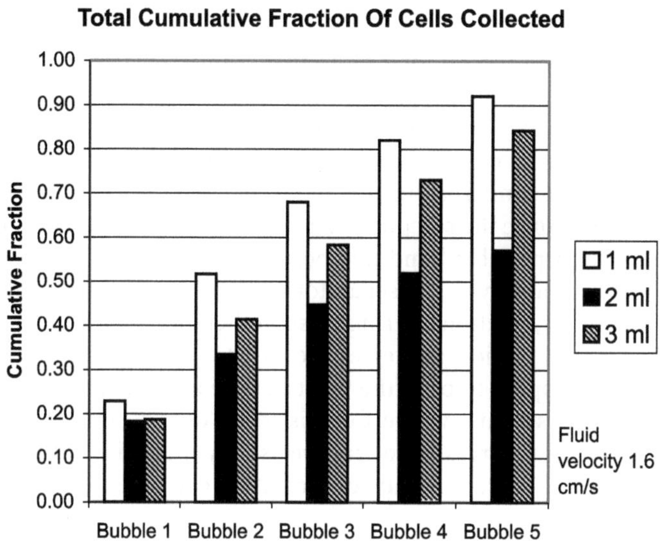

Fig. 14 Total cumulative fraction affinity adsorbed erythrocytes detached in a tubular adsorber as a function of number of bubbles passed over the surface at three bubble sizes (constant flow velocity) (from [50])

5 Conclusions

This report has focused on a number of phenomena that are important in determining the extent and strength of the affinity attachment of cells to surfaces. This requires a departure from the typical analysis used for soluble systems in that ligand density effects have been considered largely in terms of their effect on the number of possible interactions rather than in terms of liquid phase concentrations. Similarly, the relationship between apparent dissociation constants for multivalent interactions and the intrinsic dissociation constant is considered with respect to the number of interactions. The implication is that above a critical value increases in ligand density would be expected to have little influence on attachment, as is observed in practice [51].

Once multiple bonds have formed the theory presented suggests that the intrinsic dissociation constant for the ligand receptor interaction chosen is less important than the valency of the interaction in determining the stability of attachment. What is important is the speed at which these bonds can both form and break and it is interesting to note that the selectin molecules, which mediate multivalent interactions between neutrophils and the walls of blood vessels, are characterised by large dissociation constants and an abnormally high "off" constant [52]. Taking a lesson from nature it would appear that the ideal interaction to use in a cell separation should have a dissociation constant of the order of 10^{-4} mol L^{-1}, i.e. probably two orders of magnitude higher than would be typical for a molecular affinity separation. An interesting possibility for exploiting rapid detachment seen with selectin-mediated adhesion is reported by Greenberg and Hammer [53] who showed that differential rolling velocities could be used to selectively enrich sub-populations of sialyl Lewis(x)-coated microspheres.

More generally, once the receptor interaction to be used has been identified and the maximum possible bond number determined, the critical factor for success will be the ability of the separator to allow generation of the optimal chemical and flow conditions during wash and elution stages. Where chemical or biochemical elution strategies are to be employed membrane-based adsorbers would appear to offer the most flexible approach to meeting these requirements. For systems where force is to be used to effect detachment, bubble-induced shear forces are attractive as they minimise elution volumes to the point where separation and concentration can potentially be achieved in a single step [47].

Acknowledgements I would like to thank Professor Robert Eisenthal for his constructive comments and the UK BBSRC for their financial support of my work in this area.

References

1. Dean PDG, Johnson WS, Middle FA (eds) (1985) Affinity chromatography: a practical approach. IRL, Oxford
2. Graves DJ, Wu Y-T (1974) Method Enzymol 34:140
3. Cozens-Roberts C, Quinn JA, Lauffenburger DA (1990) Biophys J 58:857
4. Bell GI (1978) Science 200:618
5. Hammer DA, Apte SM (1992) Biophys J 63:35
6. Chang K-C, Hammer DA (1996) Langmuir 12:2271
7. Tees DF, Chang K-C, Rodgers SD, Hammer DA (2002) Ind Eng Chem Res 41:486
8. Krasik EF, Hammer DA (2004) Biophys J 87:2919
9. Chen S, Alon R, Fuhlbrigge R, Springer TA (1996) Prog Biophys Mol Biol 65:143
10. Olsvik O, Popovic T, Skerve E, Cudjoe KS, Hornes E, Ugelstad J, Uhlen M (1994) Crit Microbiol Rev 7:43
11. Schmitz B, Radbruch A, Kummel T, Wickenhauser C, Korb H, Hansmann ML, Thiele J, Fischer R (1994) Eur J Haematol 52:267
12. Hubble J (1997) Trends Biotechnol 15:249
13. Orsello CE, Lauffenburger DA, Hammer DA (2001) TIB Tech 19:310
14. Yang J, Bos R, Poortinga A, Wit PJ, Belder GF, Busscher HJ (1999) Langmuir 15:4671
15. Cozens-Roberts C, Quinn JA, Lauffenburger DA (1990) Biophys J 58:107
16. Lawrence MB, Springer TA (1991) Cell 65:859
17. Ming F, Whish WJDW, Eisenthal R, Hubble J (2000) Enzyme Microb Tech 26:216
18. Lam A, Cao X, Eisenthal R, Hubble J (2001) Enzyme Microb Tech 29:28
19. McAuliffe JC, Hindsgaul O (1997) Carbohydrate drugs an ongoing challenge. Chemistry and Industry, 3 March 1997, Society of the Chemical Industry, London, pp 170–174

20. Mammen M, Choi S-K, Whitesides GM (1998) Angew Chem Int Ed 37:2754
21. Hubble J (2003) Chem Eng Sci 58:4465
22. Busscher HJ, Vandermei HC (1995) Method Enzymol 253:455
23. Belfort GA, Davis RH, Zydney AL (1994) J Membr Sci 96:1
24. Hubble J, Eisenthal R, Whish WJD (1995) Biochem J 311:917
25. Hubble J, Fang M, Eisenthal R, Whish W (1996) J Theor Biol 182:169
26. Gillespie DT (1976) J Comp Phys 22:403
27. Saterbak A, Kuo SC, Lauffenburger DA (1993) Biophys J 65:243
28. MacGlashan D Jr, Lichtenstein LM (1983) J Immunol 130:2330
29. MacGlashan D Jr, Mogowski M, Lichtenstein LM (1983) J Immunol 130:2337
30. Seagrave J-C, Deanin GG, Martin JC, Davis BH, Oliver JM (1987) Cytometry 8:287
31. Golstein B, Wofsy C (1996) Immunol Today 17:77
32. Terpe K (2003) Appl Microbiol Biotechnol 60:523
33. Marsh JC, Sutherland DR, Davidson J, Mellors A, Keating A (1992) Leukemia 6:926
34. Lee YC (1992) FASEB J 6:3193
35. Biessen EAL, Beuting DM, Roelen HCPF, van de Marel GA, van Boom JH, van Berkel TJC (1995) J Med Chem 38:1538
36. Yoshida T, Toyama-Sorimachi N, Miyasaka M, Lee YC (1994) Biochem Biophys Res Comm 204:969
37. Mandrusov E, Houng A, Klein E, Leonard EF (1995) Biotechnol Prog 11:208
38. Cozens-Roberts C, Quinn JA, Lauffenburger DA (1990) Biophys J 58:857
39. Leenaars AFM, O'Brien SBG (1989) Philips J Res 44:183
40. Noordmans J, Wit PJ, van der Mei HC, Busscher HJ (1997) J Adhes Sci Technol 11:957
41. Gomez-Suarez C, Noordmans J, van der Mei HC, Busscher HJ (1999) Langmuir 15:5123
42. Gomez-Suarez C, Busscher HJ, van der Mei HC (2001) Appl Environ Microbiol 67:2531
43. Sharma SK, Mahendroo PP (1980) J Chromatogr 184:471
44. Chase HA (1994) Trends Biotechnol 12:296
45. Clemmitt RH, Chase HA (2003) Biotechnol Bioeng 82:506
46. Ujam LB, Clemmitt RH, Clarke SA, Brooks RA, Rushton N, Chase HA (2003) Biotechnol Bioeng 83:554
47. Krause S, Kroner KH, Deckwer WD (1991) Biotechnol Tech 5:199
48. Millward HR, Bellhouse BJ, Nicholson AM, Beeton S, Jenkins N, Knowles CJ (1994) Biotechnol Bioeng 43:899
49. Cao X, Eisenthal R, Hubble J (2002) Enzyme Microb Tech 31:153
50. Barkley S, Johnson H, Eisenthal R, Hubble J (2004) Biotechnol Appl Biochem 40:145
51. Ward MD, Dembo M, Hammer DA (1995) Ann Biomed Eng 23:322
52. van der Merwe PA, Barclay AN (1994) Trends Biochem Sci 19:354
53. Greenberg AW, Hammer DA (2001) Biotechnol Bioeng 73:111

Adv Biochem Engin/Biotechnol (2007) 106: 101–127
DOI 10.1007/10_2006_044

Published online: 9 June 2007

Chromatography of Living Cells Using Supermacroporous Hydrogels, *Cryogels*

Maria B. Dainiak[1,2] · Igor Yu. Galaev[1] · Ashok Kumar[2,3] · Fatima M. Plieva[1,2] · Bo Mattiasson[1] (✉)

[1]Department of Biotechnology, Center for Chemistry and Chemical Engineering, Lund University, P.O. Box 124, 22100 Lund, Sweden
Bo.Mattiasson@biotek.lu.se

[2]Protista Biotechnology AB, IDEON, 22370 Lund, Sweden

[3]Department of Biological Sciences and Bioengineering, Indian Institute of Technology, 208016 Kanpur, India

Abstract The preparative cell separation is an intrinsic requirement of various diagnostic, biotechnological and biomedical applications. Affinity chromatography is a promising technique for cell separation and is based on the interaction between a cell surface receptor and an immobilised ligand. Most of the currently available matrices have pore size smaller than the size of the cells and are not suitable for cell chromatography due to column clogging. Another problem encountered in chromatographic separation of cells is a difficulty to elute bound cells from affinity surfaces. Application of novel adsorbents, supermacroporous monolithic cryogels, allows overcoming these problems. Cryogels are characterised by highly interconnected large (10–100 μm) pores, sponge-like morphology and high elasticity. They are easily derivatised with any ligand of choice. Convective migration can be used to transport the cells through the matrix. Target cells bind to affinity ligands, while other cells pass through the cryogel column non-retained and are removed during a washing step. Because of the spongy and elastic nature of the cryogel matrices, the cells are efficiently desorbed by mechanical compression of cryogels, which provides high cell viability and yields. The release of affinity bound cells by mechanical compression of a cryogel monolithic adsorbent is a unique and efficient way of cell detachment. This detachment strategy and the continuous macroporous structure make cryogels very attractive for application in cell separation chromatography.

Keywords Supermacroporous cryogels · monolithic chromatography · cell separation · affinity chromatography · convective flow · mechanical compression

1 Introduction

Efficient and inexpensive techniques for fractionation and isolation of target cell types are necessary to provide pure cell populations for diagnostics, biotechnological and biomedical applications. Despite the growing need for methods for separation of cells into cell sub-populations, none of the existing techniques can provide a desired degree of performance at a preparative scale. The traditional techniques, micro- and ultra-filtration and ultracentrifugation which exploit differences in cell size, shape and density [1–4] have low specificity and are difficult to scale up. Flow cytometry, where the target cells are labelled with an immunofluorescent probe, is accurate but time consuming and expensive [5]. Magnetic bead separation technology [6–8] has gained popularity within the fields of cell biology and medical microbiology but is restricted mainly to analytical scale and has the problem of nonspecific cell-surface interactions and irreversible cell adsorption. Partitioning in aqueous two-phase systems [9], based on specific partitioning of cells between two immiscible aqueous polymer solutions, is a technique which has good prospects for scale up but it suffers from the necessity to separate the target product from the phase-forming polymer. Moreover, extraction in aqueous two-phase systems provides only one equilibrium stage for adsorption/desorption process and hence requires repetitive extractions in order to achieve sufficient selectivity.

Chromatography is one of the most powerful and widely used separation techniques in down stream processing of biomolecules. Adoption of this method for the separation of different types of cells offers many advantages with respect to resolution and scale-up. However, cell chromatography has not yet found wide application mainly due to the absence of adsorbents that could meet the requirements for efficient chromatographic cell separation in which the large size of cells, their low diffusivity and complex surface structure and chemistry should be considered. Traditional adsorbents for protein chromatography are not suitable for processing particulate containing feed stocks as they act as filters accumulating the cells between the beads. Macroporous beaded adsorbents [10–12] are also far from being ideal for efficient chromatographic cell separations due to diffusional limitations. Up to 95% of the convective flow passes through the void volume around beads and cells bind mainly to the surface of the beads, practically not penetrating into the pores compromising significantly the capacity of the adsorbent [13].

While expanded bed chromatography (EBC), is a promising preparative-scale technique for processing cell-containing suspensions, it suffers from some major limitations. The long equilibrium time, requirements for special columns and the narrow range of flow velocity are some of the disadvantages of EBC when it comes to specific adsorption/desorption of cells. Furthermore, high shear forces existing in EBC can affect adversely the viability of cells. Thus, a rare example of EBC application for microbial cell separation (*Escherichia coli* cells from *Saccharomyces cerevisiae* cells) is limited to the separation of heat-treated cells [14].

An alternative way is the development of macroporous chromatographic matrices with 10–100 μm large and interconnected pores, allowing micrometer size particles like cells to pass through the monolithic columns nonretained. Moreover, in monolithic columns, the mass-transport of the particles dispersed in the feed is predominantly due to convection. Thus, two main limitations of using chromatography for cell separation, namely blocking of the column by particulate material and low diffusivity of bioparticles, are circumvented when using monolithic macroporous columns.

2 Macroporous Hydrogels, *Cryogels*

It is desirable to have a process producing hydrophilic macroporous material where the voids in the material are formed by "displacing" polymer chains from the pores into pore walls, so that the pore walls have high local polymer concentration as compared to the local polymer concentration in hydrophilic nonmacroporous gels with the same formal bulk concentration of the polymer. The increased local polymer concentration in the pore walls will ensure sufficient mechanical stability of the material.

Cryotropic gelation (see review [15]) has been used in order to produce macroporous hydrophilic gels (so-called *cryogels* from the Greek *kryos* meaning *frost* or *ice*) for chromatographic applications. Though purely from the polymer science point-of-view the cryogel-type materials have existed for some time but their potential for biotechnological applications has been realized only recently [16]. Thus, the hydrophilic cryogels were specifically designed for application to technologically challenging bioseparation processes and high-throughput analysis.

Cryogels are produced via gelation processes at subzero temperatures when most of the solvent is frozen while the dissolved substances (monomers or polymer precursors of a cryogel) are concentrated in small nonfrozen regions, where the gel-formation proceeds. After melting the solvent crystals (ice in the case of aqueous media) that perform like porogen, a system of large continuous interconnected pores is formed, that provides channels for the mobile phase to flow through (Fig. 1). The pore size depends on the initial concentration of precursors in solution, their physicochemical properties and the freezing conditions (Fig. 2A,B).

It is not straightforward to determine absolute values of pore sizes in porous hydrogels as water in the pores is essential for maintaining the hydrogel integrity. The removal of water from the pores when drying cryogel samples for scanning electron microscopy (SEM) could change pronouncedly

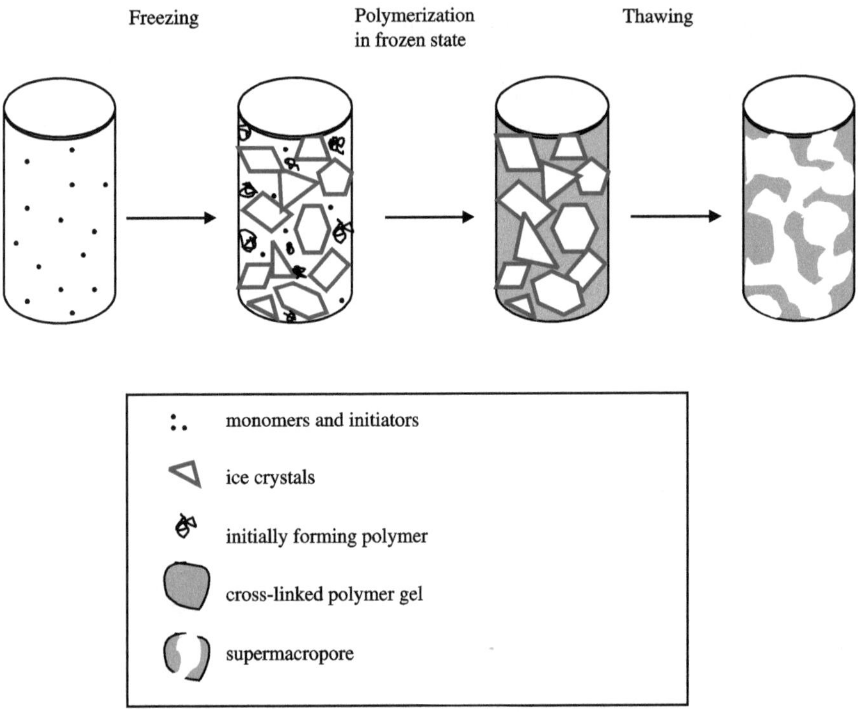

Fig. 1 Schematic presentation of the cryogel formation. Reproduced from [19]

the cryogel structure. Environmental scanning electron microscopy (ESEM) allows for the possibility of monitoring changes in the structure of the material whilst the sample dehydrates slowly [17]. In the fully hydrated sample it was impossible to visualize the pores as they were filled with water (Fig. 2C). As the water evaporated, the surface details became more visible (Fig. 2D–E). The ESEM micrographs at high degrees of dehydration showed the same macroporous structure as SEM did with interconnected pores of ten-to-hundred micrometers in size and thin and dense pore walls. Moreover, confocal microscopy of the cryogel in the wet state (Fig. 2F) demonstrated clearly in a good agreement with SEM and ESEM, the presence of large, ten-to-hundred μm interconnected pores with smooth pore walls and dense polymer walls in between the pores.

Specifically, cryogels differ from traditional gel materials due to the system of interconnected macropores (giving the cryogels a sponge-like morphology) and due to the structure of pore walls formed when compulsory increase in polymer concentration in nonfrozen regions takes place (providing cryogels with a relatively higher mechanical strength as compared to traditional gels with the same formal bulk concentration of the polymer). The concentrated polymer gel in the pore walls swells poorly and ensures elas-

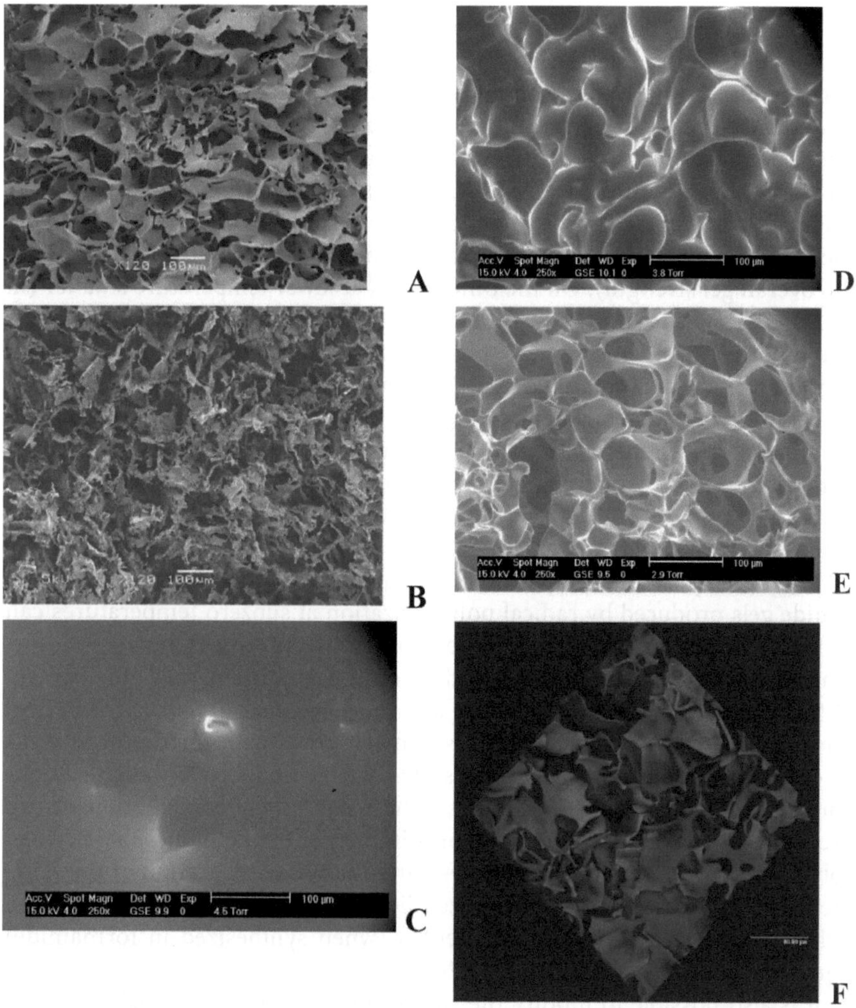

Fig. 2 Pore structure of cryogels. SEM microphotograph of diametrical cross-sections of polyacrylamide cryogels prepared at – 12 (**A**) and – 18 °C (**B**), respectively. Reproduced from [69]. ESEM images of polyacrylamide cryogel: (**C–E**) – initially wet sample at different degrees of the dehydration. Reproduced from [18]. Confocal microscopy image of fluoresceinamine-labelled polyacrylamide cryogel (**F**). Reproduced from [30]

ticity of cryogels [even those produced from such hydrophilic polymers as poly(acrylamide) and agarose], while spongy morphology allows liquid to be pumped through cryogels at high flow rates with minimal flow resistance. The structure and properties of such cryogels depend on the concentration of gel precursors and thermal conditions of cryotropic gel-formation. For instance, an increase in initial co-monomer concentration gives rise (under identical conditions) to an increase in both polymer concentration in pore walls and the overall strength of the macroporous gel material, but at the same time results in a decrease in the flow rate through the cryogel column [18]. The latter is due to the lower amount of solvent, which is frozen out from the more concentrated initial solution and, as a consequence, there is a smaller total volume of ice crystals forming the macropores. The influence of gel-formation temperature on physical properties of resulting cryogels is rather evident: the lower the freezing temperature, the higher the cryoconcentration (hence, the higher the polymer concentration in the pore walls and the higher the overall gel strength). On the other hand, at lower temperature smaller (in size) ice crystals are formed, so the smaller is the cross-section area of each macropore, and, therefore, flow resistance increases, and flow rate through a monolithic cryogel column drops [19]. The polymer concentration in the gel phase in cryogels is considerably higher as compared to conventional polyacrylamide gel, prepared from the same initial monomer solution at room temperature, thus explaining clearly why cryogels are stronger than the gels with comparable formal bulk concentration of the polymer, in spite of the macroporous morphology of cryogels.

In general, pore size, shape and morphology in the macroporous polyacrylamide gels produced by radical polymerization at subzero temperatures can be controlled in a rational way by controlling two processes, namely gel formation and solvent freezing. Depending on the conditions used (the content of the initiator, freezing temperature and the solvent used) a broad variety of porous structures ranging from closed macropores with microporous walls to open interconnected macropore systems with nonporous walls. Presumably, the bimodal pore distribution is the result of the combination of two processes, the formation of macropores in the nonfrozen liquid phase (when solvent crystals perform as a porogen) and the phase separation of the polymer synthesized within the nonfrozen liquid phase. Cryogels with uniform (when synthesized in water) or oriented (when synthesized in formamide) porosity have also been produced. [20].

It is also possible to modify the surface of pores inside cryogels by grafting with different functional polymers. The grafting to the surface could be achieved via chemical bonding between reactive groups on the gel surface and reactive terminal groups of the preformed polymer (so-called *grafting to*). The obvious advantage here is that one can beforehand determine the properties (molecular mass and molecular mass distribution) of the to-be-grafted polymer. The problem is that the surface should have reactive groups suitable for grafting and the grafted chain should carry the proper functionality at the end. It is very difficult to achieve high grafting densities using the *grafting to* methods because of steric crowding of reactive sites at the gel surface by already bound polymer molecules. Moreover, the efficiency of *grafting to* methods is pretty low resulting in pronounced losses of the terminally modified polymer. Hence, an alternative approach also called *grafting from* has been adopted for the production of polymer brushes at the pore surface. Anion-exchange polymer chains of poly(2-(dimethylamino)ethyl methacrylate) and poly([2-(methacryloyloxy)ethyl]-trimethylammonium chloride) [21,22] and cation-exchange polymer chains of polyacrylate [23] have been grafted onto poly(acrylamide)-based cryogels using potassium diperiodatocuprate as the initiator. The graft polymerization did not alter the macroporous structure of the cryogels, however, the flow rate of solutes through the cryogel matrix decreased with increase in the density of grafted polymer. The sorption of low-molecular-weight (metal ion, dye) and high-molecular-weight (protein) substances on the grafted monolithic cryogel columns indicates that a "tentacle"-type binding of protein to grafted polymer depends on the architecture of the grafted polymer layer and takes place after a certain degree of grafting has been reached [24]. The polymer brushes carrying affinity ligands attached to the polymer backbone are expected to be surfaces allowing for affinity binding of cells via multipoint interactions. On the other hand, affinity binding of cells to polymer brushes prevents the nonspecific interactions with the matrix surface and hence allows for better control of cell binding and release [25].

Unlike traditional polyacrylamide gels, which are rather brittle, polyacrylamide-based cryogels are elastic, soft, sponge-like materials that can withstand large deformations and can be easily compressed 4- to 6-fold without being mechanically damaged. The compressed monoliths re-swell and adopt their initial shape upon the addition of more liquid [26]. It is possible to compress cryogel monoliths (0.5 ml; 7.1 mm i.d.) slightly for inserting them in the wells (1.5 ml; 7.0 mm i.d.) of a standard 96-well plate [26,27] or to stack them as "building blocks" on top of each other inside the column when scale-up is needed [28,29]. Cryogel monoliths in a column format (5 ml, 12.4 mm i.d. and 4.5 ml, 7.1 mm i.d) and in a 96-minicolumn plate format were used for cell chromatography.

3
Chromatography of Bioparticles and Microbial Cells

The 10–100 μm pores (and even larger) in cryogels make it possible to use these materials for chromatography of bioparticles having sizes up to 1–10 μm without a risk of mechanical entrapment of bioparticles in the column. Even cells as big as red blood cells (7 μm) are convectionally transported by the liquid flow through the monolithic plain cryogel column without being trapped mechanically (Fig. 3A). No pronounced tailing of the cell peak was observed. One could expect the flow to be still laminar in the interconnected pores of ten-to-hundred micrometers in size at these flow rates. Hence the cells do not experience large shear forces and are intact at the outlet of the cryogel column. No haemolysis was observed in the cell-containing fractions (Fig. 3B).

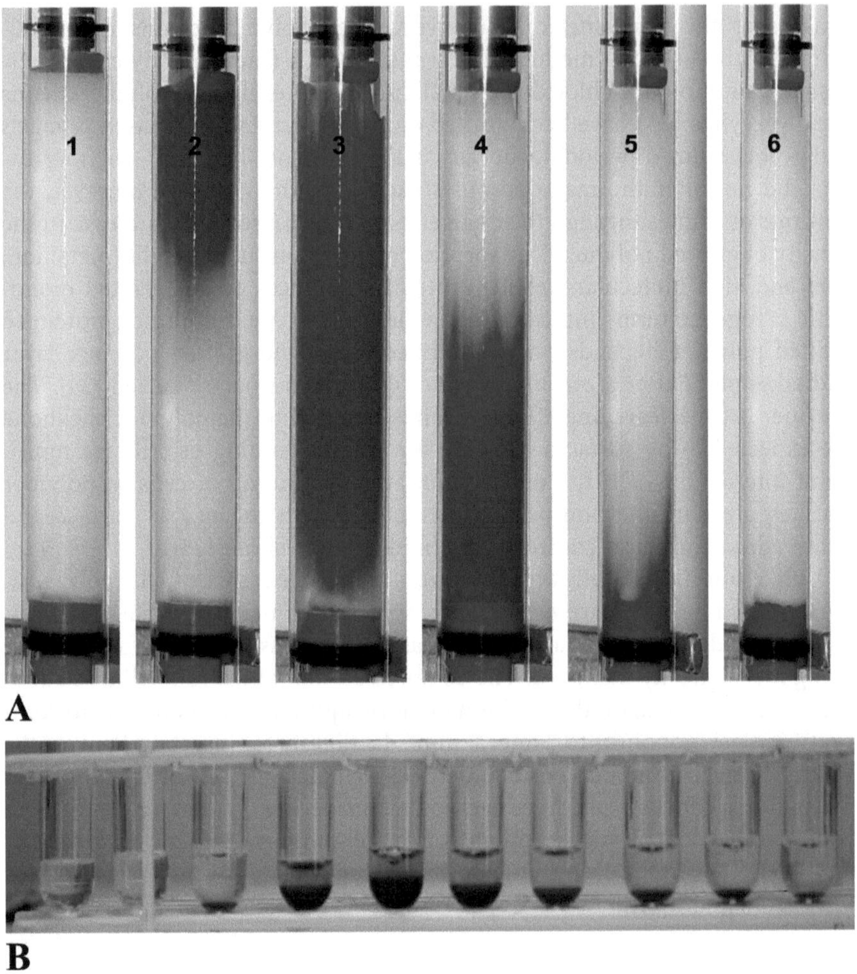

Fig. 3 Flow of the pulse of whole blood through a plain cryogel column. One ml blood was applied to the plain cryogel column at a flow rate of 0.5 ml/min in isotonic buffer solution. *column 1*: column before application; *column 2–5*: column during the run; *column 6*: column after the flow of blood sample (**A**). Flow through fractions from the column were collected and left to stand. No red blood cell lysis was observed in the fractions (**B**). Reproduced from [30]

Cells bind to the monolithic cryogel column when there is a possibility for cells to interact with some specific groups (charges, hydrophobic moieties or affinity ligands) introduced at the surface of pores in the cryogel column. Indeed, *E. coli* cells were bound to an ion-exchange monolithic cryogel column at low ionic strength and were eluted with 70–80% recovery at NaCl concentrations of 0.35–0.4 M, while *E. coli* cells bound to a cryogel column bearing Cu(II)-loaded iminodiacetate (Cu(II)-IDA) ligands were eluted with around

80% recovery using either 10 mM imidazole or 20 mM EDTA [31]. When cells are bound to the column, an important question is whether the binding takes place in the whole volume of the column or the cells are accumulated only at some particular zones, for example only at the top. One column was sacrificed after binding of ampicillin-resistant *E. coli* cells. The matrix was taken out from the column, the central disc-shaped zone was cut out and a few small pieces were taken from the central part of this disk and placed on top of an agar plate containing ampicillin (Fig. 4C). After incubation over night at 37 °C, pronounced and approximately similar growth was observed around all pieces of the matrix exposed to cells (Fig. 4A). This indicates cell binding inside the whole volume of the supermacroporous anion-exchange cryogel column. As a control, pieces of an unused column were placed on top of an agar plate without ampicillin and no growth occurred under the same

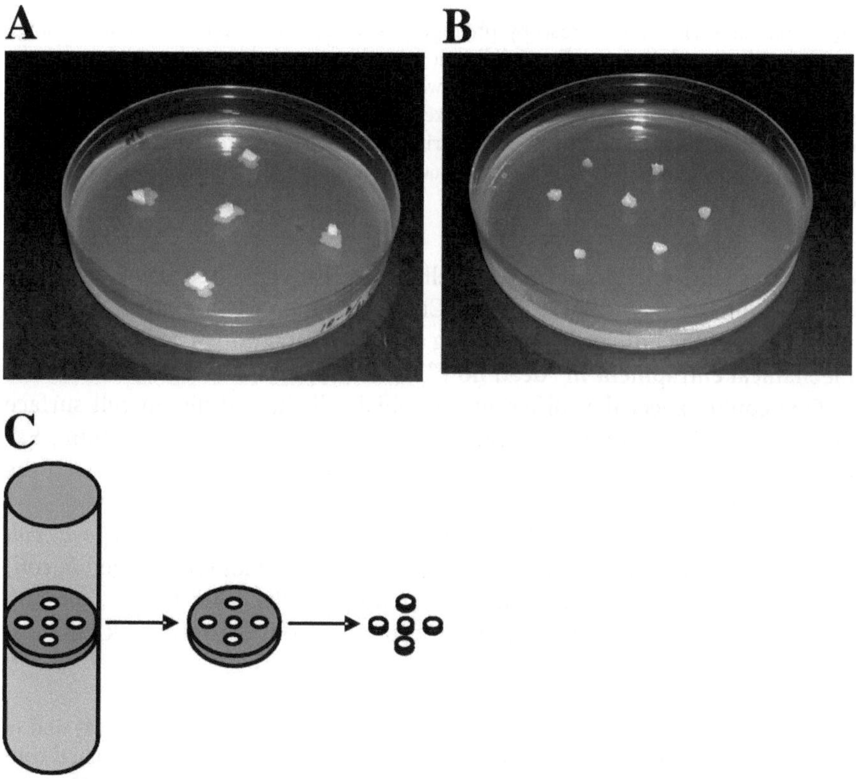

Fig. 4 **A** Growth from pieces of supermacroporous anion-exchange matrix on ampicillin-containing agar plates after binding ampicillin resistant *E. coli* cells. **B** Pieces of a newly synthesized column on an agar plate without ampicillin. Both plates were incubated for 17 h at 37 °C. **C** Schematic explanation of how pieces of the supermacroporous matrix were taken. Reproduced from [31]

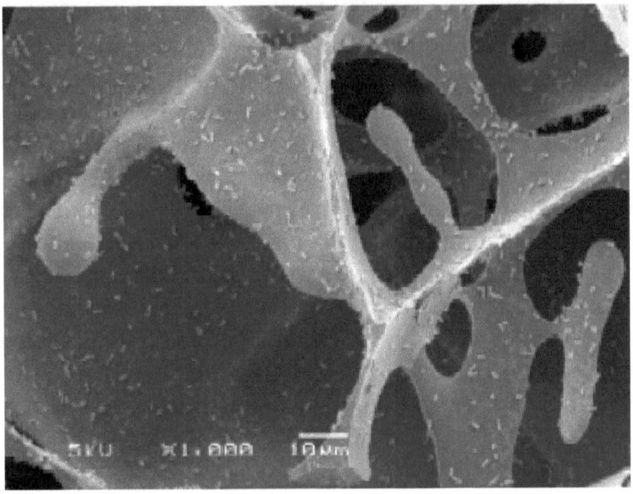

Fig. 5 Scanning electron microscopy photographs of supermacroporous anion-exchange matrix with bound *E. coli* cells at different magnification indicated by the bars at the bottom of the photographs. The samples were fixed in 2.5% glutaraldehyde in 0.15 M sodium cacodylate buffer over night, postfixed in 1% osmium tetroxide for 1 hour, dehydrated in ethanol and critical point dried. Dried samples were coated with gold/palladium (40/60) and examined using a JEOL JSM-5600LV scanning electron microscope. Reproduced from [31]

conditions (Fig. 4B). Bound *E. coli* cells were also visualized using scanning electron microscopy (SEM) (Fig. 5). Clearly, bound cells are attached to the plain parts of the pore walls indicating specific interaction of cells rather than mechanical entrapment in "dead flow" zones [31].

One could expect that different microbial cells have different cell surface properties with different chemical groups exposed to the outer medium. Exploiting these differences allows for the separation of specific cells from the mixed population.

Two model systems have been studied: the mixtures of wild-type *E. coli* and recombinant *E. coli* cells displaying poly-His peptides (His-tagged *E. coli*) and of wild-type *E. coli* and *Bacillus halodurans* cells. Wild-type *E. coli* and His-tagged *E. coli* were quantitatively captured from the feedstock contain-

Fig. 6 Chromatography profiles of wild-type *E. coli* and *B. halodurans* cells (**A**) and of their mixture (**B**) on Cu(II)-IDA cryogel column (4.5 × 1.2 cm I.D.). Experimental conditions: cell suspensions (1 ml with OD_{450} of 0.5–0.6) were passed separately (**A**) or as a mixture (**B**) (2 ml with OD_{450} of 1.0–1.2) through the column equilibrated with 20 mM HEPES, 200 mM NaCl pH 7.0 (running buffer) at a flow rate of 1 ml/min. Bound cells were eluted with 0.0–0.05 M imidazole gradient (200 ml) in the running buffer. **C** Content of wild-type *E. coli* and *B. halodurans* cells in the nonretained and retained peaks. Total amount of cells in the peak was taken as 100%. Reproduced from [32] ▶

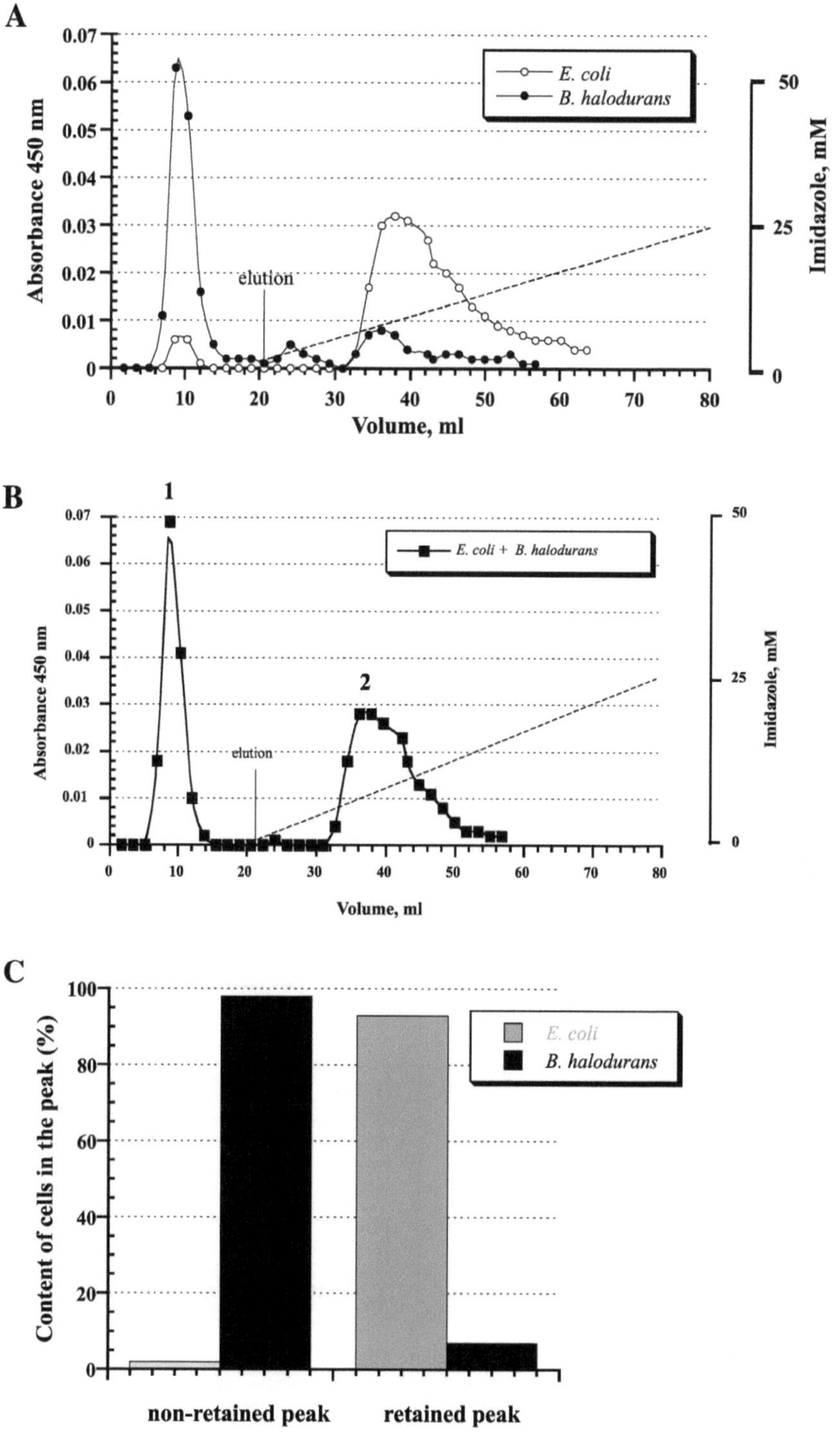
A
Absorbance 450 nm
0.07
0.06
0.05
0.04
0.03
0.02
0.01
0
E. coli
B. halodurans
elution
Imidazole, mM
50
25
0
0
10
20
30
40
50
60
70
80
Volume, ml
B
1
2
Absorbance 450 nm
E. coli + B. halodurans
elution
Imidazole, mM
Volume, ml
C
Content of cells in the peak (%)
100
80
60
40
20
0
E. coli
B. halodurans
non-retained peak
retained peak

ing equal amounts of both cell types and recovered by selective elution with imidazole and EDTA, with the yields of 80 and 77%, respectively. The peak obtained after EDTA elution was 8-fold enriched with His-tagged *E. coli* cells as compared with the peak from imidazole elution, which contained mainly weakly bound wild-type *E. coli* cells. Haloalkalophilic *Bacillus halodurans* cells had low affinity to the Cu(II)-IDA cryogel column and were efficiently separated from a mixture with wild-type *E. coli* cells, which were retained and efficiently recovered from the column with a imidazole gradient (Fig. 6). All the cells maintained their viability after the chromatographic procedure [32].

Cryogel monoliths in the format of 96 drainage-protected mini-columns proved to be very useful for a parallel assay of particulate-containing samples (Fig. 7) [26, 27]. In order to make parallel chromatography possible the plate with open-ended wells with drop-forming units at the bottom of each well was used. Studies on the binding of wild-type *E. coli* cells, recombinant *E. coli* cells with poly-His peptide displayed on the cell surface, and *B. halodurans* cells to Me(II)-IDA ligands (where Me(II) is Cu(II), Ni(II) or Zn(II)) and phenyl-cryogel monoliths in a 96-mini-column plate format revealed that both *E. coli* strains but not *B. halodurans* cells had affinity for immobilized metal ions in the following order: Cu(II) > Ni(II) > Zn(II), and wild-type *E. coli* cells had the most hydrophobic surface among the other strains studied. In contrast to recombinant *E. coli* cells, affinity of the wild-type cells

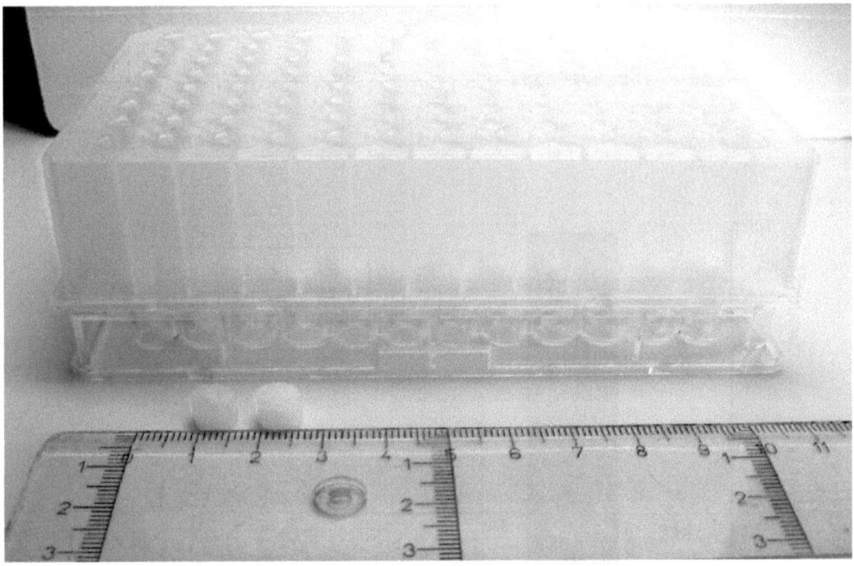

Fig. 7 Photo of the phenyl-cryogel microplate aligned on top of a 96-well clear plate and phenyl-cryogel monoliths (7.1 × 6.3 mm) removed from the phenyl-cryogel microplate. Reproduced from [35]

to immobilized metal ions significantly decreased with the prolonged fermentation time. Strongly bound cells were detached from the affinity adsorbent by mechanical compression of the cryogel monoliths. The developed screening procedure can be used for selecting ligands and optimal conditions for isolation of cells of interest from clinical and food samples [27]. The feasibility of this approach was demonstrated in the model system in which a 96-mini-column plate filled with cryogel monoliths (18.8 mm × 7.1 mm i.d.) with immobilized concanavalinA (ConA) was used for screening for suitable conditions for separating yeast and *E. coli* cells using a ConA-cryogel column (composed of six cryogel monoliths 18.8 mm × 7.1 mm stacked on top of each other) [29]. An optimization was carried out regarding the duration of cell contact with the affinity adsorbent and the load of yeast cells on the adsorbent in order to ensure quantitative capture of this type of cells during chromatographic separation. A nearly base-line chromatographic separation was achieved under optimized conditions (Fig. 8; Table 1). For the recovery of bound yeast cells a novel approach of cell detachment by mechanical compression of the elastic adsorbent was used. The flowthrough fraction

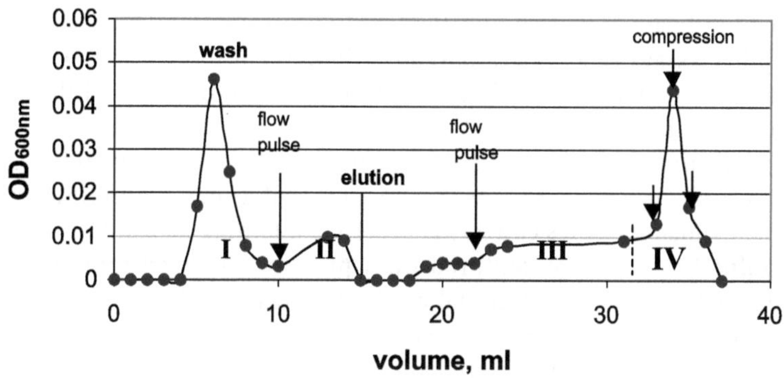

Fig. 8 Chromatogram of a mixture of *E. coli* and yeast cells on a ConA-cryogel monolithic column (112.8 mm × 7.1 mm diameter). Reproduced from [26]

Table 1 Content of *E. coli* and *S. cerevisiae* cells in the flowthrough and eluted peaks

Cell type	Cell content (%) in flowthrough (peak I[1])	in eluate (peak III[1])	in eluate (peak IV[1])
E. coli	100	11	5
S. cerevisiae	0	89	95

Total number of cells of both types in the peak was assumed to be 100%.
[1] The chromatogram is presented in Fig. 8

contained *E. coli* cells with nearly 100% purity, whereas the fraction eluted by compression of the adsorbent contained yeast cells with 95% purity. Cell detachment by elastic deformation will be discussed in details in the next section.

Cryogel monoliths stay filled with the liquid due to the capillary forces inside the pores. This phenomenon combined with bulk diffusion conditions inside the large pores, allow cell metabolism and enzymatic reactions to proceed in the monoliths with bound cells. This, in turn, allows for direct biochemical and microbiological assays of the bound cells, as was demonstrated by the example of two model systems, recombinant *E. coli* cells bound to Cu(II)-IDA monoliths and yeast cells bound to ConA monoliths [27]. The former model was analyzed using tetrazolium salt XTT which is reduced to colored formazan dye in the electron transport system in respiring cells, while the latter was assayed by monitoring pH changes of the medium, occurring due to production of protons during cell metabolism. In both models bound cells were viable and maintained ability to metabolize nutrients. In the studies of integrated capture/purification of proteins from crude homogenates the possibility of measuring enzymatic activity [26] and fluorescence [33] of proteins bound to cryogel monoliths was demonstrated. The accumulated data indicate that it should be possible to carry out essentially any of the viability assays (e.g. assays based on fluorescent dyes, immunoassays or bioluminescence analysis of ATP) of cells bound to cryogel monoliths in a 96-mini-column plate format.

Apart of microbial cells, other bioparticles, like inclusion bodies, mitochondria and viruses were specifically captured using monolithic cryogel columns. Inclusion bodies of a 33 kD protein containing 306 amino acids with three sulfur bridges were captured on a protein A-cryogel monolithic column after labeling with polyclonal antibodies against 15 and 17 amino acid residues at the N and C-terminal ends of the protein, respectively [34]. Alternatively, inclusion bodies were captured using phenyl-cryogel monolithic columns in 96 well-format. A novel ELISA system has been developed for the direct (no solubilization of the inclusion bodies needed) analysis of cryogel-captured inclusion bodies [35].

Mitochondria were captured on Cu(II)-IDA-cryogel monoliths and non-specifically released proteins were washed away allowing identification of 68 proteins released specifically upon calcium stimulation, using LC-MS/MS and database searches [36].

Chemically biotinylated moloney murine leukaemia viruses were captured on streptavidin derivatized monolithic cryogel columns from particulate-containing cell culture supernatant without preclarification of the feedstock. Adsorption capacities of 2×10^5 cfu/ml of adsorbent were demonstrated (for comparison, Fractogel streptavidin provides the capacity of 3.9×10^5 cfu/ml of adsorbent). The specific titre of the fraction recovered by the elution with 0.6 mM d-biotin was increased by 425-fold. However, recoveries of less than

8% were achieved. Adsorption of nonbiotinylated viruses on the streptavidin-cryogel monolithic columns was not observed [37].

4 Detachment of Bound Particles by Elastic Deformation

Adsorption of bioparticles, especially as big as microbial cells, to affinity surfaces involves polyvalent interactions complicating greatly the recovery of the adsorbed material. Different strategies have been tested in an attempt to overcome the problem of strong multipoint attachment of cells to affinity surfaces. For example, in a membrane-based chromatographic system bound B-cells were released by transmembrane diffusion of hydrochloric acid (pH 1) into a flow of neutralizing normal saline [38]. In approaches that do not require such drastic elution conditions cell detachment was achieved by the use of ligands immobilized through cleavable bonds [39], the passage of air-liquid interfaces [40] or by using flow-induced shear forces [41]. The latter leads to a pronounced dilution of the preparation of eluted cells and also involves the risk of cell damage.

A unique property of cryogels, namely their elasticity, has been exploited for the development of a novel system for the efficient binding and release of different cells and particles [42]. Affinity-bound bioparticles as well as synthetic particles are detached from cryogels, when the latter undergoes an elastic deformation (Fig. 9).

Intuitively, it was obvious that the matrix deformation would have no effect on the release of affinity bound macromolecules, even when these macromolecules have more than one binding site like recombinant His_6-tagged tetrameric lactate dehydrogenase from *Bacillus stearothermophilus*. Indeed, the cryogel deformation had no effect on the release of lactate dehydrogenase bound to Cu(II)-IDA-carrying cryogel either in the absence or in the presence of a specific eluent. Most probably, lactate dehydrogenase despite having a few binding sites binds to a single Cu(II)-IDA ligand as the probability of location of two or more Cu(II)-IDA ligands in the position which fits exactly the location of two binding sites on the enzyme molecule is very small.

The situation is changed completely when studying larger particles with multiple binding sites at the surface. Here the probability of multipoint attachment increases dramatically. As a model of such a particle, we have used a microgel of cross-linked poly(N-isopropyl acrylamide-*co*-N-vinylimidazole). The size of spherical microgel particles is 350–400 nm, which is somewhat larger, than the size of most virus particles, but smaller than the size of bacterial cells. Because of the imidazole groups of vinyl imidazole co-monomer, the microgel particles are capable of interacting with Cu(II)-IDA ligands.

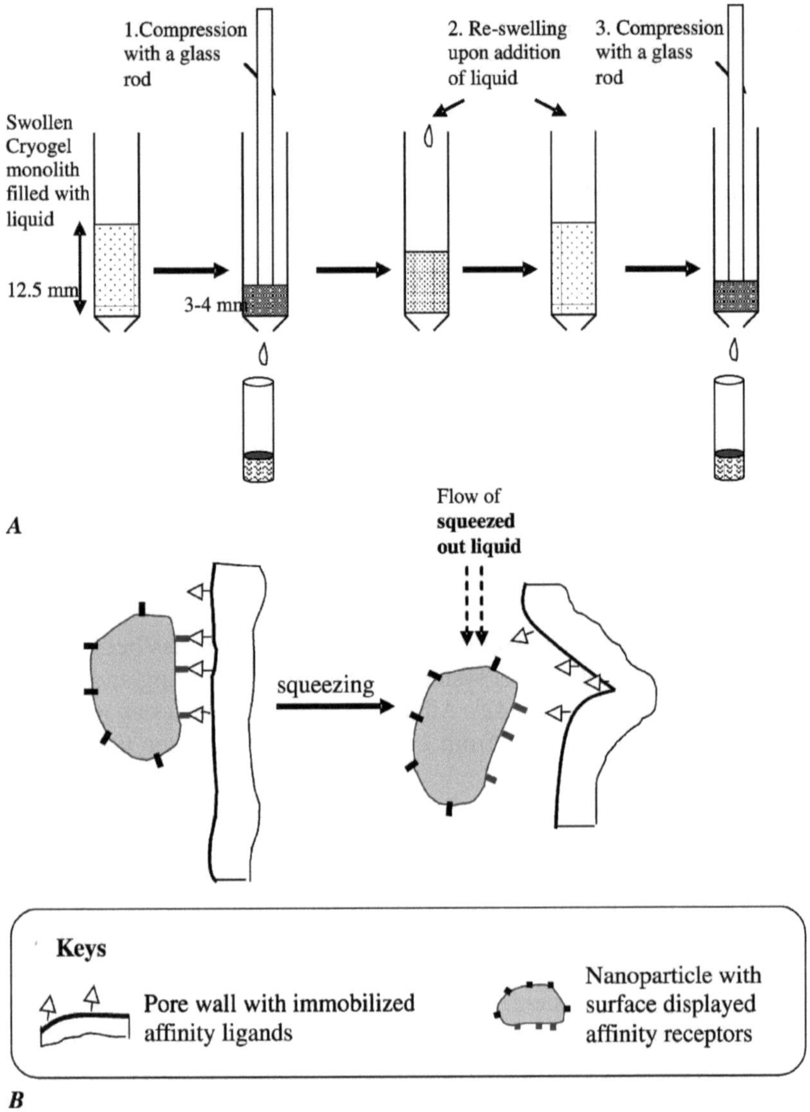

Fig. 9 Conceptual presentation of the procedure used for the release of captured particles by mechanical compression (squeezing) of monolithic affinity cryogel (**A**) and of the mechanism of detachment of captured particles induced by the squeezing (**B**). Reproduced from [42]

No bound microgel particles were eluted with running buffer or with specific eluents, imidazole (up to 0.3 M) or EDTA (up to 50 mM). However, when cryogel with bound microgel particles was compressed in the presence of either eluent, up to 60% of microgel particles was released from the cryogel. As the pulse of the eluent at 20-fold higher flow rate did not succeed in releasing more than 12% of captured microgel particles, one could assume that the primary reason for the release of bound particles is the elastic deformation of cryogel, rather than a flow generated in the pores during the MH compression.

With a size of microgel particle in the micrometer range, one could assume that multisite interactions take place between the microgel particle and the ligands at the surface of the cryogel pore. According to theoretical studies, it is unlikely that reasonable concentrations of a soluble monovalent competitor (specific eluent) can displace the binding equilibrium when the number of interactions is more than 10 [43]. For bioparticles under typical chromatographic conditions ($10^{10}-10^{12}$ of ligands and receptors per cm^2 and $10^{-10}-10^{-8}$ cm^2 of contact area) the number of specific binding interactions can be between 1 and 10 000 [40]. Thus, in order to detach the bioparticle from the matrix an external force affecting the entire bioparticle is required to simultaneously disrupt multiple bonds [44]. Alternatively, the matrix could be affected by an external force to promote the detachment of bound bioparticles. An elastic deformation of the matrix could be such a force resulting in changing the distances between the ligands at the interface and hence the development of stresses on the bound particle. As the result of these stresses, the efficiency of the interaction is reduced and some of the bonds are broken down. The presence of the specific eluent prevents the re-formation of broken bonds. As a result of these events, breakage of the existing bonds and impossibility to form new ones, the particle is released from the matrix. It was previously demonstrated by scanning electron microscopy studies that cells captured by affinity cryogels are bound to the plain "flat" parts of the pore walls and are not entrapped in "dead flow" zones [31]. The possible reasons for the disruption of affinity bonds can be the deformation of the plain surface (Fig. 9). Thus, the main driving force for the compression-induced detachment of bound particles from the surface is probably the physical dislodging of cells by microscopic deformation of the surface carrying affinity ligands and the removal of dislodged particles by the flow of squeezed out liquid. The presence of specific eluent promotes the detachment by decreasing the equilibrium number of bonds and preventing re-attachment of free particles on their way down the column.

The elasticity of cryogels affects the efficiency of the release of bound particles by compression. "Dense" (6% w/v total co-monomer concentration) cryogel has thicker walls as compared with "soft" cryogel (5%) and hence a higher elastic module (0.065 MPa) as compared to the "soft" one (0.016 MPa) [42]. The 4-fold difference in elasticity had no effect on the efficiency of the release of bound microgels and inclusion bodies. However, the release of larger particles like *E. coli* cells and especially yeast cells was highly dependent on the cryogel elasticity (Fig. 10). The more elastic the cryogel, the more efficient is the detachment of microbial cells upon the deformation.

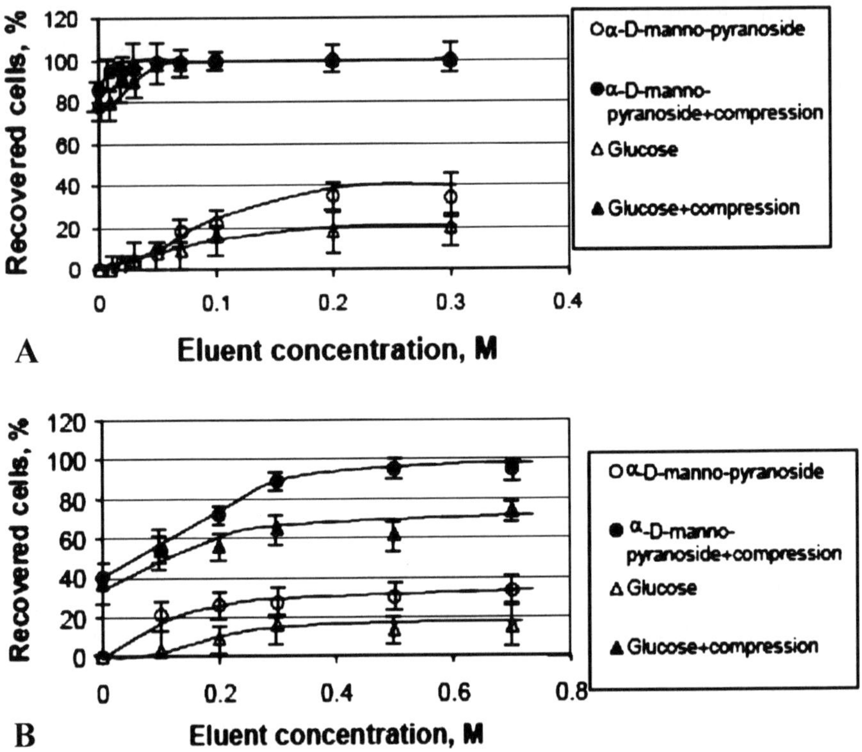

Fig. 10 Release of bound yeast cells by conventional elution and by compression of 5% (**A**) and 6% (**B**) ConA-Cryogel monoliths equilibrated with different concentrations of the eluent. The amount of bound cells was assumed to be 100%. Reproduced from [42]

It should be noticed that contrary to microgels and inclusion bodies, microbial cells are living species permanently changing their colloidal properties in response to environmental changes. For example, an incubation of yeast cells within ConA-cryogel was required for efficient capture of cells whereas the amount of captured microgel particles bound to Cu(II)-IDA-cryogel was independent of the time of contact between the applied particles and the adsorbent.

The generic nature of compression-induced detachment was demonstrated for a variety of bioparticles of different sizes like microgels (0.4 μm), inclusion bodies (around 1 μm), *E. coli* cells (1–3 μm), yeast cells (8 μm) and for different ligand–receptor pairs (IgG-protein A, sugar-ConA, metal ion-chelating ligand) [42, 45].

Apart from compression using external mechanical force, cryogels can shrink in response to the changes in the environment, provided the cryogel is made from a *stimuli responsive polymer*. Poly(N-isopropylacrylamide) (pNIPA) cryogels shrink and swell in response to temperature changes. The

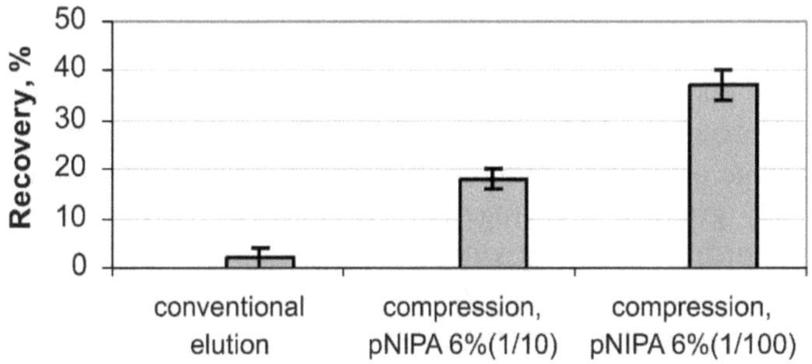

Fig. 11 Conventional elution with 0.3 M α-D-manno-pyranoside in the running buffer (0.1 M Tris-HCl with 150 mM NaCl, 5 mM $CaCl_2$, and 5 mM $MgCl_2$, pH 7.4) and temperature-induced detachment of yeast cells bound to ConA-pNIPA-macroporous hydrogels, produced from polymerization feed with total monomer concentrations 6% and MBAAm(cross-linker)/NIPA ratios 1/10 and 1/100 mol/mol, respectively. Reproduced from [45]

temperature and the amplitude of the transition depend on the cross-linking density of the gel and the total concentration of co-monomers in the polymerization feed [46].

The mechanical deformation of the pNIPA-cryogels, originated due to temperature-induced shrinkage, happened to be sufficient to facilitate the release of bound bioparticles, namely yeast cells captured via concanavalinA (ConA) covalently coupled to the pNIPA-cryogel (Fig. 11). Not surprisingly, the higher release of captured cells was observed for the pNIPA-cryogel with lower cross-linking density as the amplitude of changes was higher for this cryogel and hence one could expect higher elastic deformation of pore walls inside the cryogel monolith and as a result a stronger effect on the bound yeast cells [45].

5 Chromatography of Mammalian Cells

As chromatography on affinity supermacroporous monolithic columns proved to be a promising approach to efficient separations of individual microbial cell types, the concept was extended to the separation of specific types of mammalian cells, namely T- and B-lymphocytes.

The increasing diagnostic uses of human peripheral blood require efficient blood cell separation methods that can be used routinely to isolate different types of cells. Fractionating lymphocytes is of great importance because of the increased need for specific lymphocyte transfusion and also if one wishes to describe the roles of lymphocyte subpopulations in immunological processes. For example, T-cell enrichment techniques are vital in determining the rate of HIV progression and transplant rejection [47].

The methods available for obtaining whole populations of lymphocytes are rather straightforward, and make use of centrifugation on density gradient media, Ficoll [48]. However, to fractionate the lymphocytes into subpopulations more intrinsic techniques are required. Fractionation techniques which take advantage of physical differences between lymphocyte populations include density gradient centrifugation [49] and electrophoresis [50]. Other techniques, such as rosetting with sheep or monkey red blood cells [51], and fractionation on glass or nylon wool [52], have also been described. Most of these techniques, however, have the major disadvantage of lack of selectivity. More selective lymphocyte separation techniques such as immunomagnetic separation [53], flow cytometry [54], and immunoaffinity chromatography [55] are becoming increasingly popular and are being routinely used. Magnetic separation and flow cytometry, although representing the most powerful tools for lymphocyte cell separation, are limited to analytical applications [56].

Antibodies are highly suitable as ligands for cell separation owing to their great diversity and specificity. Separation methods for isolation of particular cell populations generally use antibodies against differentially expressed cell-surface molecules as targets. Antibodies are attached to chromatography columns and used to bind a cell that possesses an antigen recognized by the specific antibody. Distribution of cell populations in biological samples is commonly heterogeneous and most often particular cell types of interest are present in small numbers together with other major subsets. For instance, among others, the important minor cell populations include stem cells in haematological samples [57], fetal cells in maternal blood [58], residual leukemic/tumor cells from patients in clinical and morphological remission [59] or antigen-specific lymphocytes [60]. Isolation of these cell subsets in sufficient numbers with high purity and viability is commonly required in medical applications.

Because of its low cost and simple operation, cell affinity chromatography could be an attractive choice for preparative scale separation [62]. As was mentioned before, cell affinity chromatography relies on the interaction of cell surface-bound molecules and their complementary ligands (monoclonal antibodies or lectins). However, cell chromatography is different from traditional protein chromatography and entails great difficulties. As separation objects, cells are relatively large and are rather fragile and sensitive to shear stress. Their diffusivity is negligible and only convective transport can be used. Moreover, because of the multipoint attachment of the cells to the ligands on the adsorption matrix, their recovery in viable form poses some problems. Thus, for cell affinity chromatography the main requirement is the design of a suitable matrix which can be used successfully for the cell separation by addressing the above challenges. One way to circumvent low diffusivity of cells is to use expanded bed chromatography in which the stationary phase is in a form of a fluidized semi-stationary bed [14, 62, 63].

Immuno-affinity expanded bed adsorption was applied for isolation of monocytes from human peripheral blood [63]. The presence of specific receptors on the surface of the monocytes has been exploited for cell separation based on the formation of an avidin-biotin complex between biotinylated mAb-labelled cells and an avidin perfluorocarbon affinity media. The drawback of the method is that only the surface of the beads is available for cell binding and high shear rates in the expanded bed column could be detrimental for cell viability. Besides, this technique does not overcome the problem of recovering cells bound to the affinity surface. In the published protocol of the monocyte isolation, the detachment of bound cells was achieved by removing the adsorbent from the column and agitating slurry in order to elute bound cells using mechanical shear, the procedure resulting in reduced viability of the cells [63].

As a promising solution to these problems, a cryogel affinity adsorbent has been developed based on the interaction of immobilized protein A with cells bearing IgG antibodies on the surface. After treating lymphocytes with goat anti-human IgG(H + L), the IgG-positive B-lymphocytes were bound to cryogel matrix with covalently attached protein A through the Fc region of antibodies. Nonbound T-lymphocytes passed through the column (Fig. 12). More than 90% of the B-lymphocytes were retained in the column while the cells in the breakthrough fraction were enriched in T-lymphocytes (81%) (Fig. 13). The viability of the T-lymphocytes isolated was greater than 90%. About 60–70% of the bound B-lymphocytes were recovered by the addition of a displacer, human or dog IgG which competes for binding to protein A. The recovered B-cells maintained their viability [64]. The technique can be applied in general to cell separation systems provided IgG antibodies against specific cell surface markers are available. A similar approach has been employed for capture (Fig. 14) and release of human acute myeloid leukaemia KG-1 cells expressing the CD34 surface antigen [65]. The CD34 surface antigen is recognized as an important marker for haematopoietic stem cells. Thus, the system may be a good model for the separation of CD34+ cells from bone marrow or peripheral blood.

The effect of matrix architecture and ligand coupling chemistry on affinity binding of mammalian cells was shown by the capture of KG-1 cells on polyvinyl alcohol (PVA)-cryogel beads (diameter 200–500 μm) and cryogel

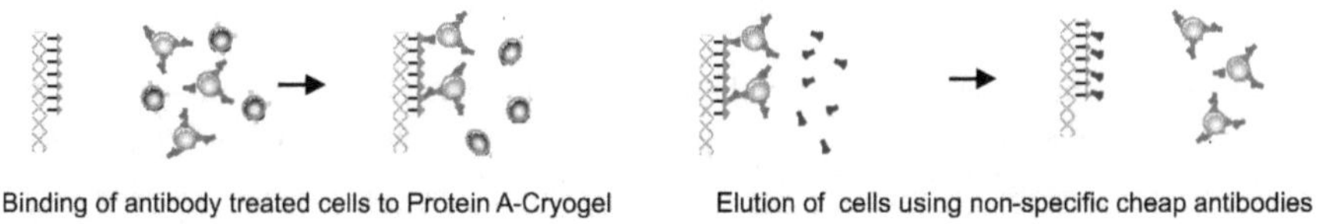

Fig. 12 Schematic presentation of separation of T- and B-lymphocytes using cryogel matrix with covalently bound protein A

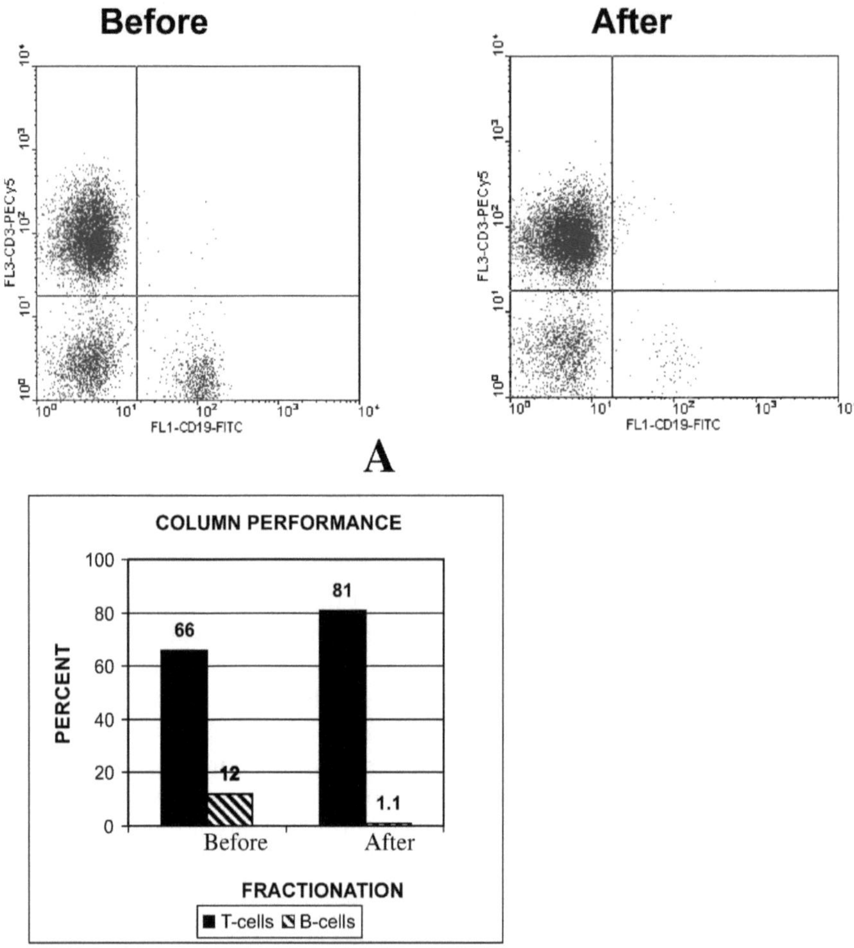

B-lymphocytes	T-lymphocytes
Binding: 91%	***Breakthrough: 81%***
Recovery: 70%	Viability: >90%
Viability: >80%	

B

Fig. 13 Composition of human peripheral blood lymphocytes before and after passage through the supermacroporous monolithic cryogel-protein A column. Lymphocytes (1 ml, 3.0×10^7 cells/ml) were treated with goat anti-human IgG(H + L) and applied to a 2 ml cryogel-protein A column. Flow cytometric analysis (**A**). Column performance (**B**). The scatter gates were set on the lymphocyte fraction. The cells bound on the column were released with 2 ml of dog IgG (30 mg/ml). Reproduced from [64]

Fig. 14 Scanning electron micrograph of the bound $CD34^+$ human acute myeloid leukemia cells (KG-1) in the inner part of supermacroporous protein A-cryogel monolithic matrix. The cells are affinity bound on the surface of the bead and on the pore walls of the monolith through the interaction of protein A and the anti-CD34 antibodies labeled on the cell surface. Magnification ×750. Reproduced from [65]

monolith with affinity ligands covalently coupled either directly to the reactive groups of the adsorbent or via a spacer [65]. About 95% of anti-CD34 labelled cells were bound to the protein A-cryogel monoliths while, only around 76% of the cell binding was achieved in the case of protein A-cryogel beads (Table 2). Surface architecture may be an important aspect, since in the monolith column the pores will form the concave surface, which presumably helps in multivalent interactions as compared with the convex surface of beads. It is important to note that, a spacer arm of seven carbon atoms was needed to improve the selective cell binding to affinity cryogel monolith while, introduction of a spacer between the ligand and the bead surface had no effect on the cell binding.

Affinity binding was lower (50%) when nonlabelled cells were applied on the cryogel beads with immobilized anti-CD34 antibodies (Table 2) probably due to the poor orientation and partial inactivation of the antibody molecules when coupled directly to the matrix. One of the most common methods of optimization of orientation of antibody molecules is by their immobilization at the Fc region through the use of protein A. This technique leaves the variable heavy and light chain regions (Fab) of immobilized antibodies available for binding to antigen epitopes [66–68]. The efficiency of cell binding to

Table 2 Specific cell capture of $CD34^+$ human acute myeloid leukemia KG-1 cells

Affinity adsorbent	Mode of cell loading	Ligand density (mg/ml adsorbent)	Binding (%)
Cryogel monolith (control)	Cells added directly	0	< 10
Protein A-cryogel monolith	Cells added directly	0.25	< 10
Protein A-cryogel monolith	Cells added after labelling with anti-CD34 monoclonal antibodies	0.25	95
CryoPVA beads (control)	Cells added directly	0	< 10
Protein A-cryogel beads	Cells added after labelling with anti-CD34 monoclonal antibodies	0.36	76
AntiCD34-cryogel beads	Cells added directly	1.23	50
AntiCD34-protein A-cryogel beads (coupling through protein A)	Cells added directly	0.4	66

Cell number: 1×10^7 cells/ml adsorbent

affinity cryogel beads was improved (66%) when anti-CD34 antibodies were indirectly immobilized through protein A.

Mammalian cells specifically bound to affinity cryogel monoliths can be recovered by mechanical compression of the adsorbent. The efficiency of de-

Table 3 Recovery of antibody-labelled CD34 human acute myeloid leukaemia (KG-1) cells from protein A-Cryogel monoliths and protein A-PVA beads

	Protein A-PVA beads			Protein A-Cryogel monoliths		
	Recovery (%)	Viability (~%)	Viable cells recovered (%)	Recovery (%)	Viability (~%)	Viable cells (%)
Elution with IgG	–	–	–	40	90	**36**
Elution with squeezing	–	–	–	75	85	**64**
Elution with IgG + squeezing	–	–	–	85	80	**68**
Elution with vortexing	60	60	**36**	–	–	–
Elution with vortexing + IgG	80	70	**56**	–	–	–

tachment and retained viability of cells eluted by compression was compared with that of cells detached by shear force, i.e. when vortexing the beads with bound cells. Compression of cryogel monoliths significantly increased the recovery of cells (Table 3). Moreover, both the efficiency of cell detachment by compression and the viability of the detached cells were significantly higher than when vortexing beads with cells bound to the bead surface [42].

6 Conclusion and Outlook

Macroporous hydrogels produced in the semi-frozen reaction media with ice crystals performing as a porogen, so-called *cryogels*, have a unique combination of properties, namely, interconnected pores of 10–100 μm size, porosity exceeding 90%, reasonably high mechanical and chemical stability, elasticity, drying and re-hydration without impairing pore structure, and retention of large amounts of liquid in the cryogel due to capillary forces. The cryogels were demonstrated to provide an efficient platform for chromatographic separation of bioparticles like inclusion bodies, mitochondria, viruses, microbial and mammalian cells, based on specific adsorption of the bioparticles. Elasticity of cryogels allows for high recovery of bound cells via elastic deformation of the cryogel under mild conditions thus ensuring retained viability of the cells.

References

1. Pertoft H, Laurent TC (1987) Separation of cells and other particles by centrifugation in colloidal silica solutions. In: Raanby B (ed) Phys Chem Colloids Macromol, Proc Int Symp. Blackwell, Oxford, p 75
2. Amici C (1991) Haematologia 70:89
3. Pretlow TG, Pretlow TP (1987) Cell Separation—Methods and Selective Applications. Academic Press, New York
4. Patel D, Ford TC, Rickwood D (1998) Fractionation of cells by sedimentation methods. In: Rickwood F (ed) Cell separation: a practical approach. Oxford University Press, New York
5. Hoffman RA, Houck DW (1997) Cell Separation using flow cytometric cell sorting. In: Recktenwald D, Radbruch A (eds) Cell separation methods and applications. Marcel Dekker Inc., New York, p 237
6. Grunn B (1991) Immunobiology 83:374
7. Bjerke T (1993) J Immunol Method 157:49
8. Sonti SV, Bose A (1995) J Colloid Interface Sci 170:575
9. Albertson P-Å (1986) Partition of Cell Particles and Macromolecules. Wiley, New York
10. Hertz CM, Graves DJ, Lauffenburger DA (1985) Biotechnol Bioeng 27:603
11. Afeyan NB, Gordon NF, Mazsaroff I, Varady L, Fulton SP, Yang YB, Regnier FE (1990) J Chromatogr 519:1

12. Regnier FE (1991) Nature 350:634
13. Rodrigues AE, Lu ZP, Loureiro JM, Gata G (1993) J Chromatogr 653:189
14. Ujam LB, Clemmitt RH, Chase HA (2002) Bioprocess Eng 23:245
15. Lozinsky VI (2002) Russian Chem Rev (Engl Edn) 71:489
16. Lozinsky VI, Galaev IY, Plieva FM, Savina IN, Jungvid H, Mattiasson B (2003) Trends Biotechnol 21:445
17. Rizzieri R, Baker FS, Donald AM (2003) Polymer 44:5927
18. Plieva FM, Karlsson M, Aguilar M-R, Gomez D, Mikhalovsky S, Galaev IY (2005) Soft Matter 1:303
19. Plieva FM, Andersson J, Galaev IY, Mattiasson B (2004) J Sep Sci 27:828
20. Plieva F, Huiting X, Galaev IY, Bergenståhl B, Mattiasson B (2006) J Mater Chem (in press)
21. Savina IN, Galaev IY, Mattiasson B (2005) J Chromatogr A 1092:199
22. Savina IN, Mattiasson B, Galaev IY (2006) J Polym Sci Part A: Polym Chem 44:1952
23. Savina IN, Galaev IY, Mattiasson B (2005) Polymer 46:9596
24. Savina IN, Galaev IY, Mattiasson B (2006) J Mol Recogn 19:313
25. Ivanov AE, Panahi HA, Kuzimenkova MV, Galaev IY, Mattiasson B (2006) Chem Eur J 12:7204
26. Galaev IY, Dainiak MB, Plieva FM, Hatti-Kaul R, Mattiasson B (2005) J Chromatogr A 1065:169
27. Dainiak MB, Galaev IY, Mattiasson B (2006) Enzyme Microb Technol (in press)
28. Dainiak MB, Kumar A, Plieva F, Galaev IY, Mattiasson B (2004) J Chromatogr A 1045:93
29. Dainiak MB, Galaev IY, Mattiasson B (2006) J Chromatogr A 1123:145
30. Noppe W, Plieva FM, Vanhoorelbeke K, Deckmyn H, Tuncel M, Tuncel A, Galaev IY, Mattiasson B (2006) manuscript
31. Arvidsson P, Plieva FM, Savina IN, Lozinsky VI, Fexby S, Bülow L, Galaev IY, Mattiasson B (2003) J Chromatogr A 977:27
32. Dainiak MB, Plieva FM, Galaev IY, Hatti-Kaul R, Mattiasson B (2005) Biotechnol Prog 21:644
33. Hanora A, Bernaudat F, Plieva FM, Dainiak MB, Bülow L, Galaev IY, Mattiasson B (2005) J Chromatogr A 1087:38
34. Ahlqvist J, Kumar A, Ledung E, Sundström H, Hörnsten G, Mattiasson B (2006) J Biotechnol 122:216
35. Ahlqvist J, Dainiak MB, Kumar A, Hörnsten EG, Galaev IY, Mattiasson B (2006) Anal Biochem 354:229
36. Teilum M, Hansson MJ, Dainiak MB, Månsson R, Surve S, Elmer E, Önnerfjord P, Mattiasson G (2006) Anal Biochem 348:209
37. Williams SL, Eccleston ME, Slater NKH (2005) Biotechnol Bioeng 89:783
38. Mandrusov E, Houng A, Klein E, Leonard EF (1995) Biotechnol Prog 11:208
39. Bonnafous JC, Dornand J, Favero J, Sizes M, Boschetti E, Mani JC (1983) J Immunol Method 11:93
40. Cao X, Eisenthal R, Hubble J (2002) Enzyme Microb Technol 31:153
41. Ming F, Whish WJD, Hubble J (1998) Enzyme Microb Technol 22:94
42. Dainiak MB, Kumar A, Galaev IY, Mattiasson B (2006) Proc Nat Acad Sci 103:849
43. Hubble J (1997) Immunol Today 18:305
44. Bell GI (1978) Science 200:618
45. Galaev IY, Dainiak MB, Plieva F, Mattiasson B (2006) Langmuir (in press)
46. Péreza P, Plieva F, Gallardo A, Roman JS, Galaev IY, Mattiasson B (2006) manuscript
47. Collins DP, Luebering BJ, Shaut DM (1998) Cytometry 33:249

48. Boyum A (1968) Scand J Clin Lab Invest 21(Suppl 97):77
49. Leise EM, Morita TN, Gray I, LeSane F (1970) Biochem Med 4:327
50. Platsoucas CD, Good RA, Gupta S (1980) Cell Immunol 51:238
51. Zola H (1977) J Immunol Method 18:387
52. Greaves MF, Brown G (1974) J Immunol Method 112:420
53. Witzig TE, Gonchoroff NJ, Ahman GA, Katzmann JA, Greipp PR (1991) J Immunol Method 144:253
54. Vral A, Louagie H, Thierens H, Philippe J, Cornelissen M, deRidder L (1988) J Radiat Biol 73:549
55. Ghetie V, Mota G, Sjöquist J (1978) J Immunol Method 21:133
56. Patel D, Rubbi CP, Rickwood D (1995) Clin Chim Acta 240:187
57. Storms et al. (1999)
58. Bianchi et al. (1990)
59. Szczepanski et al. (2001)
60. Gamadia LE, Remmerswaal EBM, Weel JF, Bemelman F, van Lier RAW, Ten Berge IGM (2003) Blood 101:2686
61. Putnam DD, Namasivayam V, Burns MA (2003) Biotechnol Bioeng 81:650
62. Clemmitt RH, Chase HA (2002) Biotechnol Bioeng 82:506
63. Ujam LB, Clemmitt RH, Clarke SA, Brooks RA, Rushton N, Chase HA (2003) Biotechnol Bioeng 83:544
64. Kumar A, Plieva FM, Galaev IY, Mattiasson B (2003) J Immunol Method 283:185
65. Kumar A, Rodríguez-Caballero A, Plieva FM, Galaev IY, Nandakumar KS, Kamihira M, Holmdahl R, Orfao A, Mattiasson B (2003) J Mol Recogn 18:84
66. Anderson GP, Jacoby MA, Ligler FS, King KD (1997) Biosens Bioelectron 12:329
67. Turkova J (1999) J Chromatogr B 722:11
68. Babacan S, Pivarnik P, Letcher S, Rand AG (2000) Biosens Bioelectron 15:615
69. Plieva FM, Savina IN, Deraz S, Andersson J, Galaev IY, Mattiasson B (2004) J Chromatogr B 807:129

Adv Biochem Engin/Biotechnol (2007) 106: 129–150
DOI 10.1007/10_2006_041

Published online: 16 December 2006

Hollow-Fibre Affinity Cell Separation

Robert E. Nordon[1] (✉) · Scott Craig[2]

[1]Graduate School of Biomedical Engineering, University of New South Wales, Samuels Building (F25), 2052 Sydney, Australia
r.nordon@unsw.edu.au

[2]Molecular Bioscience Division, John Curtin School of Medical Research, The Australian National University, Building 54, ACT 0200 Canberra, Australia

Abstract The developing fields of cell and tissue engineering will require cost-effective technologies for delivery of cells to patients. Hollow-fibre affinity cell separation is a monoclonal antibody-based cell separation process whereby monoclonal antibody (ligand) is immobilised onto a stationary substrate, the luminal surface of a parallel array of hollow fibres. Deposited cells are fractionated on the basis of adhesion strength using hollow fibre geometry that generates a well-defined shear stress for cell recovery. In this chapter we present the biophysical basis for the process of ligand-mediated cell adhesion and relate this to the performance of affinity cell separation. We also discuss the hydrodynamics of hollow fibre arrays and the various approaches for modifying polymer substrates with protein ligands. One of the major limiting factors for large-scale epitope selective cell separation will be the prohibitive cost of these affinity processes. Hollow fibre systems offer the promise of providing flexibility and scalability for many of these applications.

Keywords Cell adhesion · Cell separation · Chimeric proteins · Hollow fibres

Abbreviations

CBD Cellulose binding domain
SEM Standard error of the mean
moAb Monoclonal antibody

Dimensional quantities

a Sphere radius
C Number of receptor–ligand complexes
E_{neg} Number of epitope negative cells after selection
E_{pos} Number of epitope positive cells after selection
f_B Force per bond
F_T Total bond force ($f_B C$)
k_B Boltzmann constant
k_b Fibre bundle permeability
k_f Hydraulic permeability
k_f^0 Forward rate constant (unstressed)
k_r Reverse rate constant
k_r^0 Reverse rate constant (unstressed)
k_τ Shear stress coupling coefficient
L Fibre length
λ Range of interaction
μ Fluid viscosity
N Receptor–ligand bond number (random variable)
N_L Number of ligand molecules
N_{neg} Number of epitope negative cells before selection
N_{pos} Number of epitope positive cells before selection
ν Kinematic viscosity
P_b Axial pressure drop along fibre
p^{on} Probability of cells remaining attached
Q Flow rate
r Tube radius
R_b Header outlet radius
R_f Fibre inner radius
R_i Header inlet radius
R_T Total number of receptor molecules
ρ Fluid density
T Temperature
t_a Surface contact time
τ_{wall} Fluid shear stress at wall
$\overline{u}$ Mean axial velocity
$\overline{u_1}$ Inlet mean axial velocity
$\overline{u_b}$ Mean axial flow at the upstream face of the bundle
$\overline{u_i}$ Mean fluid velocity at header inlet
y Axial position

Non-dimensional quantities

α Receptor free energy $\left(\frac{\lambda F_T}{k_B T R_T}\right)$
α_c Critical total bond energy
β Receptor complex free energy $(\lambda F_T / k_B TC)$
ε Enrichment factor
κ Reverse rate constant (unstressed) $\left(\frac{k_r^0}{k_f^0 N_L}\right)$
M Hydraulic permeability
θ Number of receptor–ligand complexes $\left(\frac{C}{R_T}\right)$
θ_c Critical number of receptor–ligand complexes

$\hat{P}$	Axial pressure drop along fibre
κ_{pos}	Reverse rate constant of epitope positive cells
κ_{neg}	Reverse rate constant of epitope negative cells
Re	Reynolds number
$\hat{T}$	Detachment time $\left(k_f^0 N_L t_d\right)$
$\overline{U}$	Mean axial velocity $\left(\overline{u}/\overline{u_1}\right)$
Y	Axial position
Y_2	Fibre length

1 Introduction

Novel cell-based therapies have evolved out of an understanding of the relationship between cell phenotype and function. Scaling from the laboratory bench to the clinic requires cost-effective cell separation techniques ($> 10^{10}$ cells) that are selective for cell phenotype. Monoclonal antibody-based cell selection techniques have played an enabling role in the refinement of blood stem cell transplantation and the developing fields of cellular immunotherapy, tissue engineering and regenerative medicine [1].

Hollow fibre affinity cell separation is a monoclonal antibody-based cell separation process whereby monoclonal antibody (ligand) is immobilised onto a stationary substrate, the intra-capillary surface of a parallel array of hollow fibres [2–6]. Hollow fibre modules can be designed so that flow is distributed equally between hollow fibres within the array, with fully developed Poiseuille flow, so that cells are subjected to uniform fluid shear stress at the luminal attachment interface. Deposited cells can then be fractionated on the basis of adhesion strength using a geometry that generates a well-defined shear stress for cell recovery.

This chapter will examine the biophysical principles that govern this type of cell separation and how the hollow fibre concept has been implemented in practice.

2 Physical Models of Ligand-Mediated Cell Adhesion

Cell capture by the hollow fibre substrate is related to the number of receptor–ligand bonds that are formed. This process is referred to as ligand-mediated cell adhesion. One of the first attempts to model this process was by Bell [7]. The reverse rate constant was related to bond stress, the range of interaction and temperature using the following equation:

$$k_r = k_r^0 \exp \frac{\lambda f_B}{k_B T}\,, \tag{1}$$

where k_r^0 is the unstressed reverse rate constant, λ the range of interaction of the bond, f_B the force per bond, k_B the Boltzmann constant and T the temperature. If a bond is stressed by application of a detachment force, then the expression λf_B is the reversible work done on the bond and the increase in free energy of the receptor–ligand complex. The numbers of bonds that have enough thermal energy for dissociation follow the Maxwell–Boltzmann distribution.

Hammer and Lauffenburger [8] developed a dynamic model of this process. It was assumed that for attachment the bonds are unstressed:

$$\frac{dC}{dt_a} = k_f^0 N_L (R_T - C) - k_r^0 C , \tag{2}$$

where C is the number of receptor–ligand complexes, t_a unstressed surface contact time (attachment), k_f^0 forward rate constant (unstressed), k_r^0 reverse rate constant (unstressed), R_T total number of receptor molecules and N_L number of ligand molecules. It was assumed that the number of receptor molecules was small in comparison to the number of ligand molecules ($N_L \gg R_T$).

The process of detachment by stressing bonds incorporates Bell's expression for the reverse rate constant:

$$\frac{dC}{dt_d} = k_f^0 N_L (R_T - C) - k_r^0 C \exp\left[\frac{\lambda F_T}{k_B T C}\right] , \tag{3}$$

where $F_T = f_B C$.

At steady state $\left(\frac{dC}{dt_d} = 0\right)$ the rate of bond formation is balanced by the dissociation rate:

$$0 = \left(1 - \frac{C}{R_T}\right) - \frac{k_r^0}{k_f^0 N_L} \exp\left[\frac{\lambda F_T}{k_B T C}\right] \frac{C}{R_T} . \tag{4}$$

The numbers of bonds that form (C) depend on the applied bond force (F_T). If F_T is greater than a critical detachment force, then Eq. 4 which is a convex function (Fig. 1), will have no roots, $\frac{dC}{dt_d}$ is negative, and all bonds will dissociate. Below the critical detachment force, there will be two roots. A stable equilibrium will form with the bond number equal to the upper root if the initial bond number is greater than the lower root. If the initial bond number is less than the lower root, $\frac{dC}{dt_d}$ is negative, and all bonds will dissociate. This model predicts that cell adhesion should be an all-or-none phenomenon, depending on the detachment force.

Referring to Fig. 1, the critical force and bond number for cell detachment is found at the stationary point. Transforming Eq. 4 using non-dimensional quantities:

$$\theta = \frac{C}{R_T} , \quad \hat{T} = k_f^0 N_L t_d , \quad \kappa = \frac{k_r^0}{k_f^0 N_L} \quad \text{and} \quad \alpha = \frac{\lambda F_T}{k_B T R_T} . \tag{5}$$

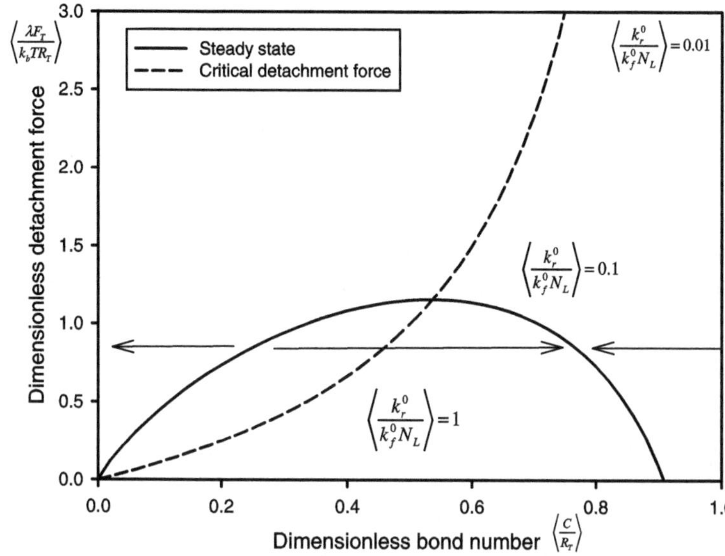

Fig. 1 Influence of detachment force on bond number (Hammer and Lauffenburger [8]) at steady state. Definition of critical detachment force

The stationary point is found by solving $\frac{\partial\theta}{\partial T} = 0$ and $\frac{\partial^2\theta}{\partial T\partial\theta} = 0$ with respect to the bond number (non-dimensional):

$$\frac{\partial\theta}{\partial\hat{T}} = (1-\theta) - \kappa\theta \exp\left(\frac{\alpha}{\theta}\right) = 0 \tag{6}$$

$$\frac{\partial^2\theta}{\partial\hat{T}\partial\theta} = \kappa\left(\frac{\alpha-\theta}{\theta}\right)\exp\left(\frac{\alpha}{\theta}\right) - 1 = 0\,.$$

Thus the critical detachment forces is related to the critical bond number by the following relation:

$$\theta_{\mathrm{c}} = \frac{\alpha_{\mathrm{c}}}{1+\alpha_{\mathrm{c}}}\,, \tag{7}$$

where the subscript c denotes critical values for the detachment force and bond number.

The cell detachment is an "all-or-none" model and does not account for the stochastic nature of fracture kinetics. Cells detach over a range of applied shear stresses [9]. Heterogeneity in cell detachment kinetics is likely related to variation in the applied forces, which depend on the shape of the cell and the method of applying a detachment force. The drag and torque applied to a sphere in contact with a plane wall in a slow uniform shear field ($Re < 1$) are directly related to the cube and square of the sphere radius, respectively [10]. So biological variation in cell shape and radius will result in even larger variations in the detachment force resulting from application of fluid shear stress. Receptor number will also vary from cell to cell, and this will be reflected by a random distribution of critical detachment forces.

Cozen-Roberts proposed a probabilistic model of cell adhesion, where stochasticity was predicted purely on the basis of small numbers of receptor–ligand bonds (< 10 000). McQuarrie has calculated the statistical fluctuations that result from reaction between small numbers of molecules using a Markov chain model [11]. Cozen-Roberts and co-workers have adopted this kinetic model to describe the stochastic nature of cell detachment phenomena [12, 13].

3 Relating Biophysical Models to Affinity Cell Separation Performance

Cell affinity techniques that utilise ligand-mediated cell adhesion rely on solid-phase immobilisation of a cell capture antibody that recognises cellular epitopes. In a cell separation process the yield of epitope positive cells will depend on the capture and release of epitope positive cells. Additionally purity is influenced by the relative abundance of epitope positive cells in the pre-separation cellular mixture and contamination of the enriched population by epitope negative cells.

Physical models of ligand-mediated cell adhesion and detachment help explain how these performance parameters are related to the physical properties of the solid phase and hydrodynamic forces.

Fluid flow will transmit viscous forces to deposited cells stressing the attachment interface. Viscous torque and drag are approximated for perfect spheres in contact with a plane wall by the following expression for low Reynolds number flow (< 1) in a linear shear field:

$$\begin{aligned} \text{drag} &= 1.7 \times 6\pi\tau_{\text{wall}} a^2 \\ \text{torque} &= 0.94 \times 8\pi\tau_{\text{wall}} a^3 \, , \end{aligned} \tag{8}$$

where a is the sphere radius and τ_{wall} is the fluid shear stress at the solid phase [10].

The relationship between the average number of receptor–ligand bonds and detachment forces (see Eq. 3) at steady state $\left(\frac{\mathrm{d}C}{\mathrm{d}t_\mathrm{d}} = 0\right)$ is expressed using dimensionless variables:

$$0 = 1 - \theta - \kappa\theta \exp[\beta] \, , \tag{9}$$

where θ is the proportion of cell receptors that are bound with ligand (C/R_T), κ is the dimensionless unstressed reverse rate constant $(k_\mathrm{r}^0/N_\mathrm{L}k_\mathrm{f}^0)$, and β is the dimensionless receptor complex bond energy $(\lambda F_\mathrm{T}/k_\mathrm{b}TC)$. The steady state equation (Eq. 9) is coupled to hydrodynamic forces by assuming that bond stress is linearly proportional to fluid shear stress (see Eq. 8):

$$\beta = k_\tau \tau_{\text{wall}} , \tag{10}$$

where k_τ is the proportionality constant.

The inherent stochastic behaviour of cell detachment processes can be related to random fluctuations in the small number of bonds that are stressed during cell detachment and variation in bond stress from cell to cell. The model of Hammer and co-workers is deterministic, predicting that all cells will detach above a critical bond stress [8]. Experimentally it is observed that cells and epitope-coated spheres detach over a range of shear stress [9, 13]. Therefore Cozen-Roberts modified this all-or-none model using Markov chain reaction kinetics [12]. We have adopted a simpler approach that assumes that the number of bonds is a Poisson random variable:

$$P[N = n] = \frac{e^{-C} C^n}{n!} , \tag{11}$$

where N, the number of bonds, is a Poisson random variable with mean equal to C.

By definition, a cell detaches when there are no bonds. So the probability of cells remaining attached when shear stress is applied is:

$$\begin{aligned} P^{\text{on}}(\tau_{\text{wall}}) &= 1 - P[N = 0] = 1 - \exp[-C] \\ &= 1 - \exp\left[\frac{-R_{\text{T}}}{1 + \kappa \exp[k_\tau \tau_{\text{wall}}]}\right] . \end{aligned} \tag{12}$$

Figure 2 shows plots of the relationship between wall shear stress and the percentage of bound cells for a range of receptor–ligand affinities. The model predicts that cells detach over a range of shear stresses and that the non-dimensional unstressed reverse rate coefficient (κ) is inversely related to the percent of cells that remain bound.

A flow cell was developed to physically characterise ligand-mediated cell adhesion inside single isolated hollow fibres [14]. A cellulose hollow fibre (external diameter 200 μm, wall thickness 7 μm), was mounted at the bottom of a 35-mm polystyrene tissue culture dish with inlet and outlet ports for filling the lumen of the hollow fibre with cells and buffer. Cells depositing inside the hollow fibre were visualised by light microscopy. The hollow-fibre module was mounted on an inverted microscope equipped with CCD camera and image analysis software. The module was connected to a simple flow system consisting of a syringe pump to control flow rate and a syringe for injecting cells into the hollow fibre.

The inner surface of the hollow fibre was modified by physical adsorption of protein, which is a chimera of protein A and a cellulose binding domain (ProtA-CBD). This adaptor molecule links domains for binding cellulose and the antibody Fc region. Cells labelled with monoclonal antibody would bind

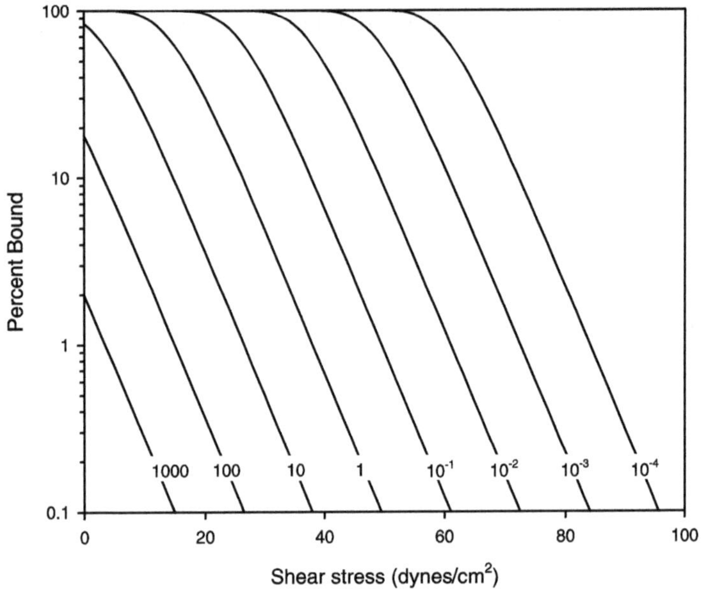

Fig. 2 Percent of cells that remain bound when shear stress is applied. The non-dimensional unstressed reverse rate coefficient (κ) is inversely related to percent bound. The simulation uses $R_T = 20$ bonds and $k_\tau = 0.2\,\mathrm{cm}^2/\mathrm{dynes}$

to the cellulose hollow fibre coated with ProtA-CBD. The direct interaction of antibody with cell surface receptors was studied by immobilising monoclonal antibody onto a hollow fibre coated with ProtA-CBD.

Cell adhesion would be measured by injecting a cell suspension (5×10^6 cells/mL) into the hollow fibre, resulting in the deposition of around 150 cells per 2 mm fibre segment. Flow was commenced following a period of deposition with zero flow. Images were captured of the deposited cells and after application of shear stress. The fraction of cells that remained bound at selected flow rates was calculated from the ratio of cell counts before and after application of shear stress.

The influence of cell deposition time and the surface density of ligand on cell adhesion were investigated by measuring the relationship between the fractions of cells that remained bound and the detachment shear stress. Figure 3a shows a plot of detachment shear stress and fraction of cells bound for deposition times of 4 min (closed circles) and 8 min (open circles). The erythroleukaemic cell line, KG1a, was labelled with antibody against the CD34 antigen. The mouse Fc region of the mouse monoclonal antibody (isotype 2a) displayed by KG1a cells bound to fibres coated with ProtA-CBD. There was stronger adhesion at 8 min compared to 4 min. Figure 3b shows cell-binding data for two different immobilised ligand densities. Fibres coated with ProtA-CBD were incubated with a 1 : 10 or 1 : 20 dilution of mouse moAb (mono-

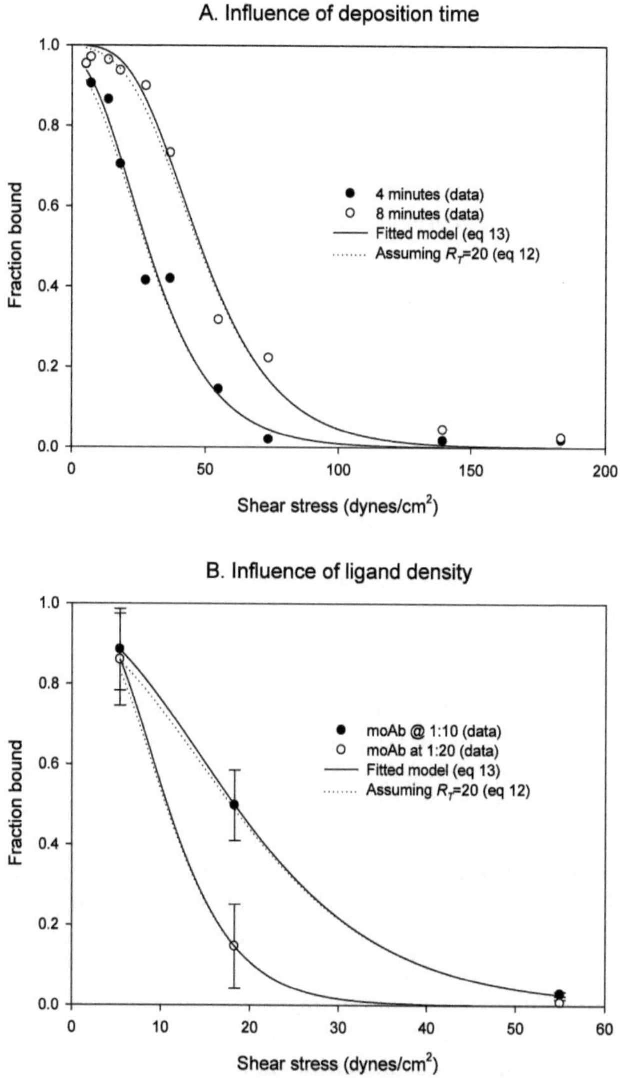

Fig. 3 **a** Influence of cell deposition time on cell adhesion: Hollow fibres were incubated overnight with ProtA-CBD (40 μg/mL). KG1a cells labelled with anti-CD34 moAb were injected into hollow fibres followed by 4 min (*closed circles*) or 8 min (*open circles*) of deposition at zero flow. Each data point represents the average of two adhesion runs. **b** Influence of ligand density on cell adhesion: ProtA-CBD coated fibres were incubated with a 1 : 10 or 1 : 20 dilution of anti-CD34 moAb. Unlabelled KG1a cells bound directly to the modified membrane. The *horizontal axis* is the recovery shear stress, and the *vertical axis* is the proportion of cells that remained bound after application of shear stress. The data was fitted to a two-parameter model of cell adhesion (Eq. 13) and the best-fit curves are displayed as *continuous lines*. The *dotted curves* correspond to the three-parameter model (Eq. 12), which includes an assumed cell receptor number term ($R_T = 20$). The best fit parameters are shown in Table 1

clonal antibody) against CD34 for 2 h before washing and attachment of KG1a cells. There was stronger adhesion when the surface was coated with the 1 : 10 dilution of mouse moAb.

Non-linear regression analysis (Sigmaplot, version 8.0, SPSS) was used to fit the model (Eq. 12) to data. It was not possible to estimate R_T and κ individually because they were linearly correlated at the shear stress values that were tested. This is not surprising since for $R_T > 20$ receptors Eq. 12 is approximated by:

$$P^{on}(\tau_{wall}) = 1 - \exp\left[\frac{-R_T}{1 + \kappa \exp\left[k_\tau \tau_{wall}\right]}\right] \quad (13)$$

$$= 1 - \exp\left[\frac{-1}{1/R_T} + A \exp[k_\tau \tau_{wall}]\right]$$

$$\approx 1 - \exp\left[\frac{-1}{A \exp[k_\tau \tau_{wall}]}\right],$$

where $A = \kappa/R_T$. So R_T and κ are linearly related by the coefficient A.

The approximate model (Eq. 13) was fitted to the data to give estimates for A and k_τ and is shown as continuous curves on Fig. 3. The two models (Eqs. 12 and 13) were in close agreement if it was assumed that R_T was at least 20 receptors (dotted curves, Fig. 3). To estimate the value of R_T, it would be necessary to measure the force required to break only a few receptor–ligand bonds. Characterisation of the adhesive force of relatively few receptor–ligand bonds (< 20) would require apparatus to measure detachment at shear stresses less than 5 dynes/cm^2.

Table 1 shows the fitted values for the two-parameter model (Eq. 13). A larger value of A could be related to a decrease in the unstressed reverse rate constant (κ) or an increase in the number of cell receptors (R_T) that are recruited to the attachment site. The geometric and mechanical properties of the cell and attachment site would influence the value of k_τ, the proportionality constant relating wall fluid shear stress to bond stress.

The estimated value of A is higher for shorter deposition times. The values for A (mean $\pm$ SE) were 0.261 ± 0.040 versus 0.118 ± 0.027 at 4 and 8 min, respectively. Longer deposition times may result in recruitment of cell receptors towards the attachment site, resulting in smaller A. Theoretically, the dimensionless unstressed reverse rate constant κ should not be influenced by deposition time. The value of k_τ did not change significantly after 4 or 8 min of contact with the membrane (0.060 ± 0.006 versus 0.052 ± 0.005 cm^2/dyne) suggesting that the geometry of the attachment site did not vary from 4 to 8 min.

The ligand density on the membrane was altered by changing the concentration of moAb in the binding reaction between immobilised ProtA-CBD and soluble moAb. The weaker adhesion at lower ligand surface density was re-

Table 1 Two-parameter model for ligand-mediated cell adhesion (see Fig. 3 and Eq. 13)

Immobilised ligand	Cell receptor	Membrane contact time (min)	A	k_τ (cm^2/dyne)
ProtA-CBD	Mouse Fc	4	0.261 ± 0.040	0.060 ± 0.006
		8	0.118 ± 0.027	0.052 ± 0.005
1 : 10 moAb	CD34 antigen	4	0.288 ± 0.092	0.088 ± 0.021
1 : 20 moAb			0.177 ± 0.072	0.195 ± 0.043

lated to a decrease in A (0.288 ± 0.092, 0.177 ± 0.072) and an increase in k_τ (0.088 ± 0.021, 0.195 ± 0.043). The enhanced adhesion at higher ligand density may be related to recruitment of cell receptors to the attachment site, increasing R_T, and to an increase in the attachment area, lowering k_τ. Thus it appears that there is some dependence between the geometry of the attachment site and the kinetics of ligand–receptor interactions.

A cell separation process is defined by assuming that the cell mix consists of two sub-populations with different κ values. Epitope positive cells (κ_{pos}) bind to the immobilised ligand, and negative cells (κ_{neg}) have non-specific affinity for the substrate. Separation efficiency is related to the ratio of these reverse rate constants.

In the cell depletion step, cells are deposited onto the substrate where cell adhesion occurs. Adherent cells are fractionated from non-adherent cells by applying fluid shear stress (washing). Epitope positive cells are depleted in the recovered washings. In a second step, epitope positive cells are enriched by recovery of bound cells by fluid shear stress and releasing agents that disrupt receptor–ligand bonds.

The cell enrichment factor (ε) is defined as:

$$\frac{N_{pos}}{N_{neg}} \times \varepsilon = \frac{E_{pos}}{E_{neg}}, \tag{14}$$

where N and E refer to the number of cells before separation and in the enriched fraction, respectively. The subscripts refer to epitope positive and negative cells. If it is assumed that all cells that remain bound to the substrate are recovered in the enriched fraction, then:

$$\varepsilon = \frac{P_{pos}^{on}(\tau_{wall})}{P_{neg}^{on}(\tau_{wall})} = \frac{1 - \exp\left[\frac{-R_T}{1+\kappa_{pos}\exp[k_\tau \tau_{wall}]}\right]}{1 - \exp\left[\frac{-R_T}{1+\kappa_{neg}\exp[k_\tau \tau_{wall}]}\right]}, \tag{15}$$

where $P_{pos}^{on}(\tau_{wall})$ and $P_{neg}^{on}(\tau_{wall})$ are the probability that epitope positive and negative cells remain bound after application of shear stress (see Eq. 11).

Figure 4 shows that a plot of enrichment factor versus shear stress is a sigmoidal function, the maximal enrichment equalling the ratio $\kappa_{neg}/\kappa_{pos}$. Thus, the enrichment factor will not exceed the relative affinity of the receptor–ligand interaction for epitope positive and negative cells. Separation performance is limited by the selectivity of the affinity substrate.

Given an affinity substrate, the washing shear stress will directly influence the enrichment purity and yield defined below. If all of the captured cells are recovered during the enrichment step then the enrichment yield will equal the probability of epitope positive cells remaining bound after the washing step:

$$\text{Enrichment yield} = P_{pos}^{on}(\tau_{wall}) = 1 - \exp\left[\frac{-R_T}{1 + \kappa_{pos}\exp[k_\tau \tau_{wall}]}\right]. \tag{16}$$

The numbers of positive and negative cells in the enriched population defines the purity:

$$\begin{aligned}\text{Enrichment purity} &= \frac{E_{pos}}{E_{pos} + E_{neg}} \\ &= \frac{N_{pos}P_{pos}^{on}(\tau_{wall})}{N_{pos}P_{pos}^{on}(\tau_{wall}) + N_{neg}P_{neg}^{on}(\tau_{wall})} \\ &= \frac{1}{1 + \frac{1}{\varepsilon} \times \frac{N_{neg}}{N_{pos}}}.\end{aligned} \tag{17}$$

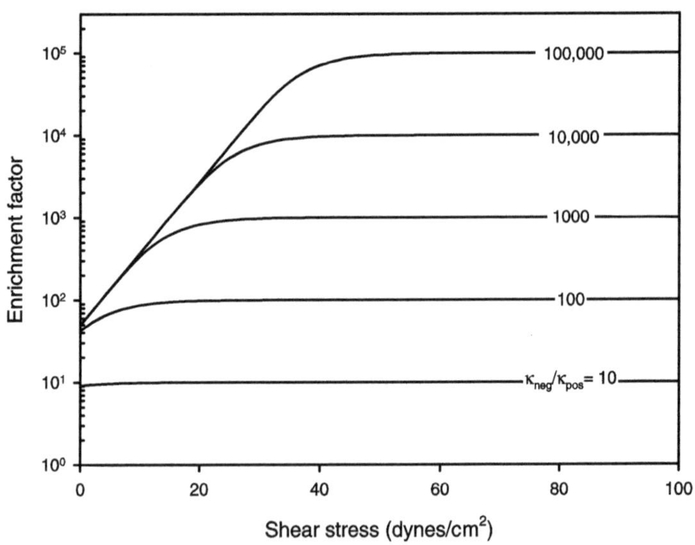

Fig. 4 Enrichment factor versus shear stress used to wash substrate. Enrichment factor is directly related to the shear stress and increases to a maximum value that is equal to $\kappa_{neg}/\kappa_{pos}$. The simulation uses $R_T = 20$ receptors, $k_\tau = 0.2\ \text{cm}^2/\text{dynes}$ and $\kappa_{neg} = 1000$

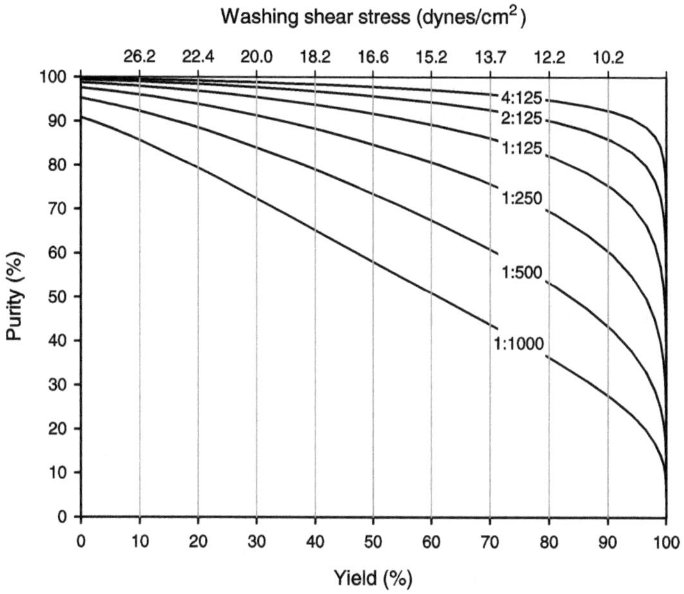

Fig. 5 Optimisation of enrichment purity and yield by selection of washing shear stress. Purity can be increased at the cost of yield. Each curve assumes a different ratio of epitope positive to negative cells before separation (range 1 : 1000–4 : 125). The simulation uses $R_T = 20$ bonds, $k_\tau = 0.2\ \mathrm{cm}^2/\mathrm{dynes}$, $\kappa_{neg} = 1000$, $\kappa_{pos} = 1$. Purity and yield are determined by the washing shear stress, which is shown on the *top axis*

The expression for purity can be expressed in terms of the enrichment factor (ε, see Eq. 14) and the ratio of positive to negative cells before separation.

The dependency of purity and yield on washing shear stress is depicted in Fig. 5. Purity has been plotted with respect to yield for an affinity substrate ($\kappa_{neg} = 1000$, $\kappa_{pos} = 1$, $R_T = 20$ bonds, $k_\tau = 0.2\ \mathrm{cm}^2/\mathrm{dynes}$). The family of curves has members with different ratios of positive to negative cells before separation (1 : 1000–4 : 125). There is an inverse relationship of yield to purity that is sensitive to the washing shear stress (see top axis).

4 Uniform Shear Elution Affinity Cell Separation

A calibrated fluid shear stress can be generated by parallel plate or tube geometries. For tube geometry, wall shear stress for fully developed laminar flow is:

$$\tau_{wall} = \frac{4Q\mu}{\pi r^3}\,, \qquad (18)$$

where Q is flow rate, μ fluid viscosity and r the tube radius.

Cell detachment forces are directly proportional to flow rate, and cells are fractionated on the basis of their adhesion strength. Purity and yield are sensitive to the fluid shear stress that is used to fractionate depleted from enriched sub-populations. Therefore, a cell separation device that generates uniform fluid shear stress at the attachment interface would improve separation performance.

5
Design of Hollow Fibre Modules

Initially we developed a small-scale parallel-plate system to study the influence of shear stress on separation performance [9], but larger surface areas could be manufactured cost-effectively using the hollow fibre spinning technology developed for renal dialysis [3, 4].

A typical renal dialyser is a hollow fibre array housed within a polycarbonate shell with two independent flow paths (intra- and extra-capillary) separated by a dialysis membrane. The intra-capillary surface area is approximately $1\ \text{m}^2$ with the theoretical possibility of binding up to 10^{10} cells/m^2, assuming that a single cell occupies a $10\ \mu\text{m}$ square.

Hollow fibre systems for affinity cell separation require the generation of uniform surface fluid shear stress at the attachment interface, the intra-capillary surface of fibres. Fibres are permeable to small molecules and water ($< 10\ \text{kD}$), so it is possible that the pressure drop along the hollow fibre bundle during flow could result in fluid leak across the membrane and a non-uniform axial flow velocity. Another hydrodynamic factor that may lead to non-uniformity in shear stress at the fibre inner wall would be a non-uniform radial distribution of flow to the hollow fibre bundle [15].

Figure 6 shows a diagram of a hollow fibre module where the extra-capillary compartment has been sealed (fixed volume). Flow along the inside of hollow fibres generates a pressure gradient and there will be flow across the membrane depending on the transmembrane pressure difference and the membrane hydraulic permeability. At steady state, the extra-capillary space will have a positive pressure, with fluid leaving fibres upstream and returning into fibres downstream (back-filtration). So flow along the fibre decreases towards the middle of the module.

The mean axial velocity is a function of axial distance:

$$\overline{U} = \cosh[MY] + B \sinh[MY]\,, \tag{19}$$

where $\overline{U} = \overline{u}/\overline{u_1}$ is the non-dimensional axial mean velocity, $\overline{u}$ is the mean axial velocity, and $\overline{u_1}$ is the inlet mean axial velocity. The non-dimensional axial position Y is defined as y/R_f where y is the axial position, and R_f is the fibre inner radius. M is the non-dimensional fibre wall hydraulic permeability

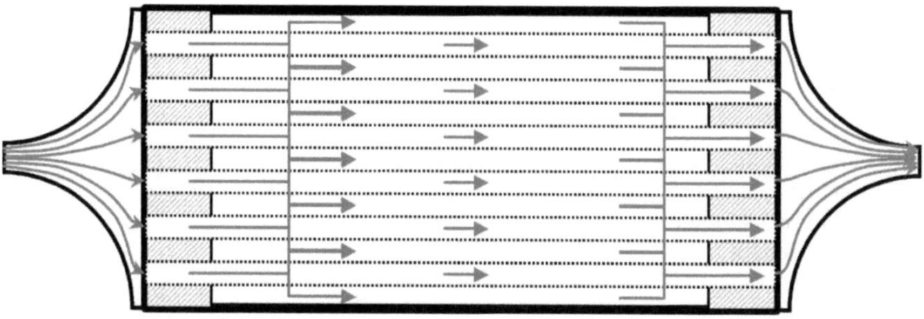

Fig. 6 Hollow fibre module hydrodynamics. The upstream header distributes flow into a bundle of semi-permeable hollow fibres. Flow along the inside of hollow fibres generates a pressure gradient and there will be flow across the membrane depending on the trans-membrane pressure difference. If the module shell is closed (fixed volume), there will be a positive pressure in the extra-capillary space, so fluid will be driven out of fibres at the upstream end, and back into fibres at the downstream end (backfiltration)

and equals $4\sqrt{k_f/R_f}$ where k_f is the hydraulic permeability:

$$B = \frac{1 - \cosh[MY_2]}{\sinh[MY_2]}, \tag{20}$$

where $Y_2 = L/R_f$ is the non-dimensional fibre length.

The minimum flow velocity will occur at $Y_2/2$. The velocity at the midpoint of the fibre for the dialyser specification shown in Table 2 was 95% of the inlet velocity ($Y_2/2 = 4907$ and $M = 6.6 \times 10^{-5}$ were substituted into Eqs. 19 and 20). Micro-porous hollow fibre systems, which have higher hydraulic permeability (e.g. $M = 0.001$) would be unsuitable for this application since there will be significant leakage of water across the membrane during flow (mean velocity at midpoint is 1.5% of inlet velocity if $M = 0.001$ and $Y_2/2 = 4907$).

It is possible that header geometry will influence the module's radial distribution of flow inside hollow fibres. Figure 6 depicts the flow in inlet (outlet) headers as diverging (converging) streamlines. It is possible that most of the flow will pass through fibres in the neighbourhood of the central axis of the module. Because the cross-sectional area of the header increases near the upstream face of the fibre bundle there will be deceleration of fluid and an increase in pressure. Close to the fibre bundle interface, the stagnation pressure at the module inlet $\left(\frac{1}{2}\rho\overline{u_i}^2\right)$ will approximate this pressure (ρ = fluid density and $\overline{u_i}$ = mean fluid velocity at header inlet).

The hollow fibre bundle is potted at each end in a solid resin, and the hollow fibre bundle acting as a resistance element dissipates the fluid inertia generated by the constricted neck of the inlet header. The permeability of the fibre bundle as a whole (k_b) can be assumed to be independent of device flow

Table 2 Dimensions for a typical renal dialyser (Polyflux 6L (low flux), Gambro)

Parameter	Value
Fibre inner diameter (R_f)	107.5 μm
Wall thickness	50 μm
No of fibres per module	9660
Active fibre length	210 mm
Surface area	$1.36\ m^2$
Ultrafiltration rate	2.9×10^{-11} m/s/Pa
Hydraulic permeability (k_f)	2.9×10^{-14} m
Non-dimension permeability ($4\sqrt{k_f}/R_f$)	6.6×10^{-5}

rate and defined using Darcy's law:

$$\overline{u_b} = \frac{k_b}{\mu} P_b \,, \tag{21}$$

where $\overline{u_b}$ is mean axial flow at the upstream face of the bundle and P_b is the pressure drop along the fibre bundle.

The permeability of the fibre bundle in the axial direction can be modelled as a parallel array of equally spaced tubes. Poiseuille's law is applied to derive an expression that relates device dimensions to fibre bundle permeability:

$$k_b = \frac{nR_f^4}{8R_b^2 L} \,, \tag{22}$$

where R_b is the radius of the bundle of fibres.

An important dimensionless quantity that predicts behaviour of the header flow pattern as defined by numerical simulations (Fluent™) is the dimensionless pressure drop parameter [15]. This parameter ($\hat{P}$) is defined as the ratio of the bundle pressure drop to the dynamic pressure at the header inlet and is also related to the ratio of the header inlet and outlet radii (R_i and R_b):

$$\hat{P} = \frac{P_b}{1/2\rho\overline{u_i}^2} = \frac{2\nu}{k_b\overline{u_b}}\left(\frac{R_i}{R_b}\right)^4 \,, \tag{23}$$

where $\nu = \mu/\rho$ is kinematic viscosity.

If $\hat{P} > 50$ then the flow pattern inside the module is predominately determined by the geometry of the hollow fibre bundle. If fibres are uniformly distributed, flow will be equally distributed to individual fibres within the bundle. In conclusion, design of hollow fibre modules for cell separation should reflect the criteria for uniform axial flow, taking into account the hydraulic permeability of hollow fibres and the effect that header design has on the radial distribution of axial flow.

6
Design of Bioactive Polymer Substrates

There is increasing interest in the development of bioactive polymer substrates for cell separation and tissue engineering applications. A biologically active surface consists of molecular domains that engage cell surface receptors either to perform cell separation or to influence cell function at a transcription level by signalling pathways that are activated by receptor engagement. Extracellular matrix molecules or cell surface receptors are usually composed of more than one functional subunit or domain, so it may be simpler to engineer proteins that are chimeras of selected functional domains.

Another approach is to immobilise functional peptides or peptidomimetics that are produced by solid phase chemical synthesis. Peptidomimetics are peptides whose amino acid sequence has no resemblance to the biological domain, but which engage cell surface receptors with similar specificity and affinity to the "wild-type" molecule.

Larger protein domains are isolated and synthesised using recombinant DNA techniques. The production process requires over-expression of the protein domain(s) in a bacterium, plant cell, yeast or mammalian cell type. Genetic domains from different species and proteins can be spliced together to create proteins with novel function, so called fusion or chimeric proteins.

Affinity cell separation techniques require relatively large surface areas that are coated with ligand. High efficiency immobilisation techniques are required because protein and peptide ligands are relatively expensive to produce. Proteins are immobilised onto polymeric surfaces by physical adsorption, covalent chemistry or hapten-mediated binding. Physical adsorption of a specific protein is difficult to predict because of its reversible nature. Other serum proteins displace physically adsorbed proteins (e.g. albumin, vitronectin, fibronectin). Covalent chemistry via protein $-NH_3$, $-SH$ or $-COOH$ side groups provides irreversible binding. Coupling may result in random orientation and chemical cross-linking of the protein, all of which lead to decreased ligand activity.

We have modified cuprophan (regenerated cellulose) hollow fibres with hydrazide groups. Monoclonal antibody was coupled to the hydrazide support by using the method originally described by O'Shannesy [16]. Sialic acid residues present on immunoglobulin carbohydrate side chains were oxidised to form aldehyde groups, which reacted with hydrazide groups to form stable hydrazone bonds [17].

We also investigated the strategy of hapten-mediated binding, which relies on the highly specific interaction between an immobilised hapten and a protein that possesses a hapten-binding domain. The hapten is a small molecule and is usually cheaper to covalently link to the polymer compared to the protein ligand. The protein ligand is bound and oriented via the hapten onto the surface of the polymer. The concentration of protein required to modify the surface can be lower than that required for covalent chemical reactions since the hapten-binding domain has micromolar to nanomolar affinity. For example, the hapten biotin is commonly used to tag proteins that will then bind to streptavidin-labelled proteins (e.g. streptavidin-phycoerythrin).

There are naturally occurring proteins that bind and degrade cellulose substrates. Cellulases are produced by bacteria and fungi to degrade cellulosic material found in the plant world. These cellulose-degrading enzymes exist as single proteins or multiprotein complexes, and consist of discrete domains exhibiting a specific affinity interaction or catalytic activity. Cellulose binding domains (CBDs) are found within the cellulases and non-enzymatic protein components of the larger cellulose-degrading complexes, and have been utilised in various biotechnology applications.

There are now many reported CBD fusion proteins with a wide range of CBD domains that have affinity for various forms of cellulose with micromolar affinity [18]. The cellulose binding domain utilised in our studies was isolated from *C. cellulovorans*. This domain was fused with the antibody-binding protein, protein LG [19, 20], to create the fusion protein CBD–LG.

Cellulose hollow fibres were coated with CBD–LG to capture antibody-labelled cells. The single hollow-fibre assay for ligand-mediated cell adhesion was used to demonstrate the functional activity of this fusion protein [14]. Figure 7 shows a photomicrograph of a $CD34^+$ cell line (KG1a cells) that was labelled with antiCD34 monoclonal antibody (IgG2a) and captured by immobilised CBD–LG.

Another problem associated with the design of bioactive polymer substrates for cell selection is the level of non-specific cell attachment. Mononu-

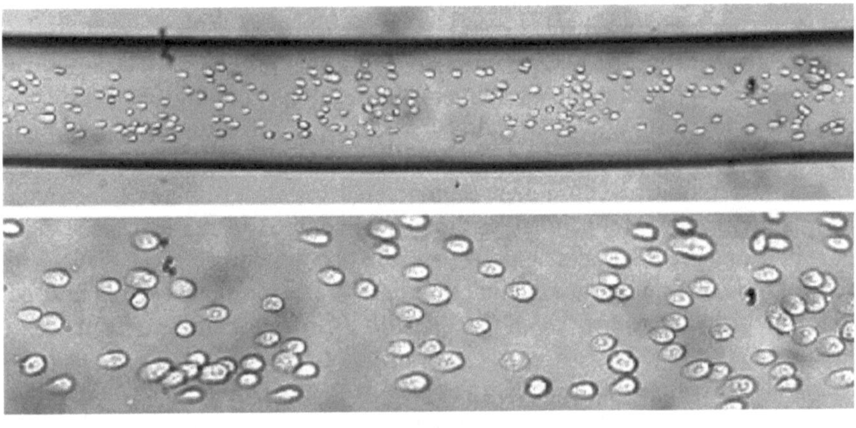

Fig. 7 Cells captured on a cellulose hollow fibre using CBD–LG. Labelled or unlabelled KG1a cells ($CD34^+$) were inject into a hollow fibre coated with various CBD ligands and incubated for 4 min before washing at defined shear stress. Bound KG1a cells were labelled with antiCD34 (IgG2a) and captured by immobilised CBD–LG. The cells elongated at higher shear stress (50 dynes/cm^2)

clear cell concentrates from blood or bone marrow will have specialised cell types that adhere to most surfaces (monocytes, platelets and granulocytes) and these cell types will contaminate enriched populations. Therefore, it is necessary to have a second round of enrichment where epitope positive cells are selectively released from the polymer substrate.

There are various approaches to the disruption of antigen-antibody bonds in a selective manner. These dissociation mechanisms include reduction of disulphide bridges [21], enzymatic proteolysis targeting specific peptide sequences [3, 4, 22, 23] and competitive binding using peptides [24] or polyclonal antibodies [25].

7
Cell Selection Process

Hope et al. published the first study describing the application of hollow fibre modules to affinity cell selection. CD4-positive lymphocytes were depleted from peripheral blood mononuclear cells [6]. Goat anti-mouse antibodies were covalently attached to cellulose hollow fibre modules using a silane-coupling reagent. Peripheral blood mononuclear cells were labelled with mouse moAb against CD4. The starting cell suspension was pumped through the dialyser followed by a wash with collection of the effluent. The process was repeated, with up to three depletion cycles. The percent depletion of $CD4^+$ cells was between 63–99.9% (two experiments). Non-$CD4^+$ cells were also depleted (36–53%) indicating that the separation process had poor specificity.

Another study by Mandrusov et al. employed a flat cellophane dialysis membrane that was sandwiched between two headers, which dual matching flow channels on either side of the membrane [2]. The cell separation channel had a height of 0.2 mm. The matching second channel on the opposite side of the membrane was for passing cell elution buffers. The cellulose membrane was activated with carbonyldiimidazole and reacted with goat anti-mouse immunoglobulin. The membrane's selective affinity was tested using mouse B-lymphocytes, which expresses mouse immunoglobulin.

The cell suspension was flowed over the modified cellulose membrane at a shear rate of $15\,s^{-1}$. This was followed by a wash at a shear rate of $315\,s^{-1}$. Cell elution was induced with pH 5 and pH 2 hydrochloric acid supplied directly to the cell side of the membrane, or pH 1 hydrochloric acid supplied by diffusion across the membrane by flow along the channel on the non-cell side of the membrane. On average, the percentage of live cells in the effluent was 30% and 10% for pH 5 and pH 2 applied to the cell side of the membrane. The viability was between 70–100% for acid pH 1 supplied directly to the opposite side of the membrane. Thus it was possible to partially protect cell from the toxic effect of low pH by utilising an acidic boundary layer approach.

Characterising cell adhesion kinetics inside hollow fibres optimised the process further. Cell adhesion is incomplete even at low shear stress so we chose a period of cell deposition without flow to maximise cell depletion. Channel heights smaller than 200 μm are required to reduce cell deposition times to less than 2 min. It is necessary to inject cells into the intra-capillary compartment of the hollow fibre module so that flow rate can be used to fractionate cells by uniform shear stress. Cells are evenly distributed along fibres using a combination of open-ended flow and blind-ended filtration [3]. Modules are aligned horizontally so that cells rapidly deposit onto the ligand-coated substrate (< 2 min). For 200 μm internal diameter fibres, cell binding will be achieved within 4–8 min (Fig. 3). If there is multilayer cell deposition, then axial rotation of the module can be used to increase the likelihood of cell contact with the ligand-coated substrate (e.g. rotation through 90° every 4 min).

A shear fractionation process replaces "washing procedures", so that increasing flow rate and shear stress elutes cells with different binding affinities. Once the epitope negative cell population has been recovered, the remaining bound epitope positive cells are released from the substrate using enzymes or competitive elution agents.

The process was tested with mobilised peripheral blood mononuclear cells from seven patients undergoing blood stem cell transplant procedures for non-Hodgkin's lymphoma or breast cancer. The CD34 antigen is a cell marker commonly used to enrich blood stem cells for human transplantation. Monoclonal antibody to the CD34 antigen was covalently coupled to the luminal surface of Cuprophan hollow fibres using hydrazide chemistry. After selective adsorption of $CD34^+$ cells (28 min), a depleted fraction was collected at 5 dynes/cm^2 followed by washes at 10 and 25 dynes/cm^2. Antigen-positive cells were recovered at 5 dynes/cm^2 after incubation with chymopapain, a protease that selectively cleaves the CD34 antigen.

The average number of cells processed was $1.3 \pm 0.2 \times 10^8$ ($\pm$SEM) and the pre-selection frequency of $CD34^+$ cells was $1.6 \pm 0.6\%$ (range 0.21–4.13%; $n = 7$). The enrichment purity was $94.4 \pm 3.1\%$, and $61 \pm 9\%$ of input CD34-positive cells was recovered in the enriched fraction ($n = 4$). Enrichment resulted in $3.3 \pm 0.1 \log_{10}$ depletion of CD34-negative cells.

8 Future Developments

The developing fields of cell and tissue engineering will require cost-effective technologies for delivery of cells to patients. One of the main limiting factors for large-scale phenotype selection is the prohibitive cost of the process. Recombinant proteins or peptidomimetics may reduce the cost of ligand production, and hapten-mediated adsorption of ligands may offer a relatively simple technique for modifying polymer surfaces. The solid phase support is regenerated by washing at high shear and then reused for multiple rounds of cell binding and recovery. Thus it may be feasible to develop economical separation processes that are suited for clinical and large-scale industrial applications. Hollow fibre systems offer the promise of providing the flexibility and scalability for many of these applications.

References

1. Nordon RE, Schindhelm K (1999) Ex vivo cell therapy. Academic Press, San Diego
2. Mandrusov E, Houng A, Klein E, Leonard EF (1995) Membrane-based cell affinity chromatography to retrieve viable cells. Biotechnol Prog 11:208
3. Nordon RE, Haylock DN, Gaudry L, Schindhelm K (1996) Hollow-fibre affinity cell separation system for CD34+ cell enrichment. Cytometry 24:340
4. Nordon RE, Milthorpe BK, Slowiacek PR, Schindhelm K (1998) Cell separation device. Unisearch Ltd, Sydney, Australia, US Patent 5 763 194
5. Orsello CE, Lauffenburger DA, Colton CK (1999) Characterization of cell detachment from hollow fiber affinity membranes. Biomed Sci Instr 35:315
6. Pope NM, Kulcinski DL, Hardwick A, Chang YA (1993) New application of silane coupling agents for covalently binding antibodies to glass and cellulose solid supports. Bioconj Chem 4:166
7. Bell GI (1978) Models for the specific adhesion of cells to cells. Science 200:618
8. Hammer DA, Lauffenburger DA (1987) A dynamic model for receptor-mediated cell adhesion to surfaces. Biophys J 52:475
9. Nordon RE, Milthorpe BK, Schindhelm K, Slowiaczek PR (1994) An experimental model of affinity cell separation. Cytometry 16:25
10. O'Neill ME (1968) A sphere in contact with a plane wall in a slow linear shear flow. Chem Eng Sci 23:1293
11. McQuarrie DA (1963) Kinetics of small systems. Int J Chem Phys 38:433
12. Cozens-Roberts CL, Lauffenburger DA, Quinn JA (1990) Receptor-mediated cell attachment and detachment kinetics. I. Probabilistic model and analysis. Biophys J 58:841
13. Cozens-Roberts CL, Lauffenburger DA, Quinn JA (1990) Receptor-mediated cell attachment and detachment kinetics. II Experimental model studies with the radial-flow detachment assay. Biophys J 58:857
14. Nordon RE, Shu A, Camacho F, Milthorpe BK (2004) Hollow fiber assay for ligand-mediated cell adhesion. Cytometry 57A:39
15. Nordon RE, Schindhelm K (1997) Design of hollow fiber modules for uniform shear elution affinity cell separation. Artif Organs 21:107
16. O'Shannessy DJ (1990) Hydrazido-derivatized supports in affinity chromatography. J Chrom A 510:13
17. O'Shannessy DJ, Dobersen MJ, Quarles RH (1984) A novel procedure for labeling immunoglobulins by conjugation to oligosaccharide moieties. Immunol Lett 8:273
18. Tomme P, Boraston A, McLean B, Kormos J, Creagh AL, Sturch K, Gilkes NR, Haynes CA, Warren RA, Kilburn DG (1998) Characterization and affinity applications of cellulose-binding domains. J Chrom B Biomed Sci Appl 715:283
19. Craig S (2003) Development of chimeric protein ligands based on domains from the Clostridium cellulosome for affinity purification. School of Biotechnology and Biomolecular Science, University of New South Wales, Sydney, p 258

20. Kihlberg BM, Sjobring U, Kastern W, Bjorck L (1992) Protein LG: a hybrid molecule with unique immunoglobulin binding properties. J Biol Chem 267:25583
21. Grimsley PG, Amos TA, Gordon MY, Greaves MF (1993) Rapid positive selection of $CD34^+$ cells using magnetic microspheres coated with monoclonal antibody QBEND/10 linked via a cleavable disulphide bond. Leukemia 7:898
22. Civin CI (1992) Release of cells from affinity matrices. The John Hopkins University, Baltimore MD, US Patent 5 081 030
23. Sutherland DR, Marsh JC, Davidson J, Baker MA, Keating A, Mellors A (1992) Differential sensitivity of CD34 epitopes to cleavage by Pasteurella haemolytica glycoprotease: implications for purification of CD34-positive progenitor cells. Exp Hem 20:590
24. Tseng-Law J, Kobori JA, Al-Abdaly F (1995) Positive and positive-negative cell selection mediated by peptide release. US Patent 5 968 753
25. Rasmussen AM, Smeland EB, Erikstein BK, Caignault L, Funderud S (1992) A new method for detachment of Dynabeads from positively selected B lymphocytes. J Imm Met 146:195

Adv Biochem Engin/Biotechnol (2007) 106: 151–171
DOI 10.1007/10_2006_045

Published online: 18 January 2007

Cell Partitioning in Aqueous Two-Phase Polymer Systems

J. M. S. Cabral

Institute for Biotechnology and Bioengineering,
Centro de Engenharia Biológica e Química,
Instituto Superior Técnico, Av. Rovisco Pais, 1049-001 Lisbon, Portugal
joaquim.cabral@ist.utl.pt

Abstract This review addresses whole cell separation and isolation using aqueous two-phase systems based on biocompatible polymers. The physicochemical factors that influence phase separation and systems properties are analysed. Especially, emphasis is given to the polyethylene glycol (PEG) and dextran two-phase systems and to stimuli-responsive soluble–insoluble polymers. The major factors that affect cell partitioning, such as polymer molecular weight and concentration, temperature, ionic species and pH, and affinity extraction, are also evaluated taking into account the cell types and cell surface properties. The applications of aqueous two-phase separation in cell processing are described, namely the new developments in continuous cell partitioning in microdevices and extractive bioconversions with relevance to the biomedical sector.

Keywords Affinity · Animal Cells · Aqueous Two-Phase Systems · Cell Separation · Partitioning

Abbreviations

ATPS	Aqueous two-phase system
C	Concentration
CF	Concentration factor
CHO	Chinese hamster ovary
EOPO	Ethylene oxide–propylene oxide copolymer
IDA	Iminodiacetate
IMA	Immobilized metal ion affinity
IMAP	Immobilized metal ion affinity partitioning
K	Partition coefficient
LCST	Lower critical solution temperature
MAb	Monoclonal antibody
MW	Molecular weight
NIPAM	*N*-Isopropylacrylamide
PEG	Polyethylene glycol
pI	Isoelectric point
PVP	Polyvinylpyrrolidone
STL	Slope of the tie line
TLL	Tie-line length
VR	Volume ratio
Y	Yield

1 Aqueous Two-Phase Polymer Systems

Aqueous two-phase partitioning of proteins, cells and cell organelles is a well-documented process which was introduced by Albertsson [1]. Detailed discussions of basic and applied aspects can be found in several monographs and reviews [1–5]. This technique is very powerful for biomaterials separation, primary downstream processing steps and extractive bioconversions. The main advantages have been summarized by Albertsson [1] and are given below:

1. Both phases of the system are of an aqueous nature.
2. Rapid mass transfer and mixing until equilibrium requires little energy input.
3. Enables the processing of solid-containing streams.
4. Polymers stabilize proteins.
5. Separation can be made selective.
6. Easy and reliable scale-up from small laboratory experiments.
7. Possibility of continuous operation.
8. It is cost effective.

The simplest procedure of this technique is the one-step extraction. The phase system is prepared and the mixture to be separated is added. After mixing, phase separation is accomplished either by settling under gravity or by centrifugation. The phases are separated and analysed or used to recover the separated components of the initial mixture. The target product (e.g. particulate, biomolecule, cells) should be concentrated in one of the phases and the contaminants in the other.

The theoretical yield in the top phase, Y_t, can be calculated in relation to the volume ratio of the phases, VR (volume top/volume bottom), and the partition coefficient K of the target biomaterial ($K = C_{top}/C_{bottom}$):

$$Y_t = \frac{100}{1 + (1/\mathrm{VR})(1/K)} \ (\%) .$$

The theoretical concentration factor in the top phase, CF_t, of a biomaterial is defined as the ratio between the target biomaterial concentration in the top phase and the target biomaterial concentration in the input mixture. This can be given as a function of the theoretical yield, volume ratio and the weight percentage of media added to the separation system:

$$\mathrm{CF}_t = \frac{Y_t}{100} \frac{\%\,\mathrm{media}}{100} (1 + 1/\mathrm{VR}) .$$

In Figs. 1 and 2, the theoretical yield and concentration factor are depicted against the volume ratio for several K values of the target and total biomaterial, and different mixture loads, respectively [6]. For high volume ratios, higher yields can be achieved. However, this is accomplished at a cost of increased dilution of the input mixture to be separated, lower purification factors and less usage of the chemicals per unit weight of mixture, which is economically unfavourable.

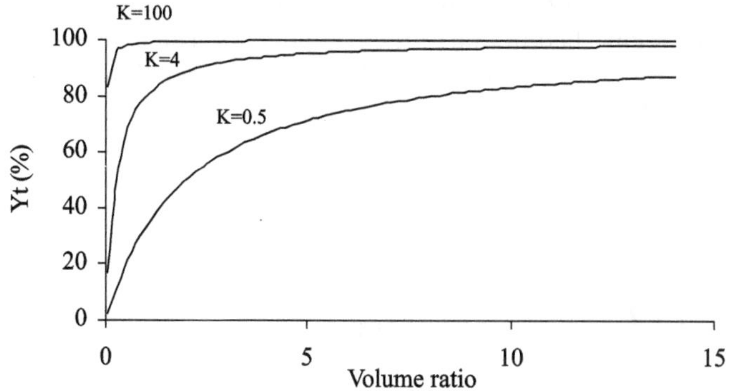

Fig. 1 Theoretical product yield in the top phase for different partition coefficients of the target biomaterial

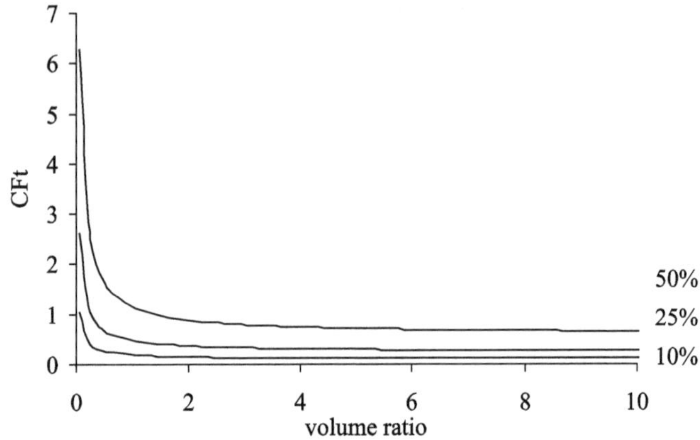

Fig. 2 Theoretical product concentration factor in the top phase for different mixture loads (weight percentage) considering a partition coefficient for the target biomaterial of 20

2
Phase Separation and System Properties

The phase components in aqueous two-phase systems (ATPSs) may be either two different hydrophilic polymers, such as polyethylene glycol (PEG) and dextran, or one polymer and one low molecular weight solute, usually a salt, such as potassium phosphate. Above certain critical concentrations of these components, phase separation occurs. Separation is dependent on the molecular weight (MW) of the polymers, additives, pH and temperature. Each of the phases is enriched in one of the components. The composition of each phase can be determined for the total system composition from the phase diagram (Fig. 3) [6].

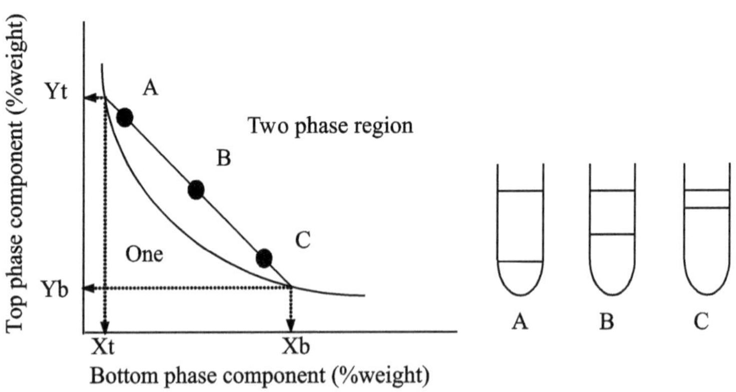

Fig. 3 Two-phase diagram for Y–X–water system

In Fig. 3, the three systems A, B and C differ in their initial compositions and in the volume ratios. However, they all have the same top phase equilibrium composition (X_t,Y_t) and the same bottom phase equilibrium composition (X_b,Y_b). This is because they are lying on the same tie-line, whose end points determine the equilibrium phase compositions and lie in a convex curve, named the binodal, which represents the separation between the two immiscible phases.

It should be noted that commercial polymers are usually polydisperse and their molecular weight distributions may vary from lot to lot, even when obtained from the same manufacturer. The phase diagrams for systems formed by different lots differ accordingly [6].

The tie-line length (TLL) and the slope of the tie-line (STL) can be related to the equilibrium phase composition as follows [6]:

$$\mathrm{TLL} = \sqrt{(X_b - X_t)^2 + (Y_t - Y_b)^2}$$

$$\mathrm{STL} = \frac{(Y_t - Y_b)}{(X_t - X_b)} = \frac{\Delta Y}{\Delta X} .$$

Table 1 shows some common examples of ATPSs. For more extensive lists, the reader is referred to the references Albertsson [1], Zaslavsky [4] and Diamond and Hsu [5].

Table 1 Polymer systems capable of phase separation in water solutions

Polymer	*Polymer*
Polyethylene glycol	Polyvinyl alcohol
	Dextran
	Hydroxypropyl starch
	Ficoll
Polypropylene glycol	Methoxypolyethylene glycol
	Polyethylene glycol
	Dextran
Ethylhydroxyethylcellulose	Dextran
	Hydroxypropyl starch
Ethylene oxide–propylene oxide	Dextran
	Hydroxypropyl starch
Polymer	*Low molecular weight solute*
Polypropylene glycol	Potassium phosphate
	Glucose
Polyethylene glycol	Inorganic salts, e.g. K^+, Na^+, Li^+, $(NH_4)^+$, PO_4^{3-}, SO_4^2
Stimuli-responsive polymer	–
Ethylene oxide–propylene oxide	–
Poly-*N*-isopropylacrylamide	–
Methacrylic acid–methyl methacrylate	–

2.1 Two-Polymer Systems

The most important factor for phase separation is the chemical nature of both polymers. In two-polymer ATPSs, the phase separation is due to small repulsive interactions between the two types of monomers in the solution. The total interaction between the two polymers is large since each one is composed of several monomers.

2.1.1 Molecular Weight

The higher the MWs of the polymers, the lower is the polymer concentration required for phase separation [1, 5], i.e. the binodal is depressed. Forciniti and co-workers [7] evaluated the effect of polymer MW and temperature in the phase composition of a PEG/dextran ATPS. The TLL, $\Delta_{\mathrm{Dextran}}$ and Δ_{PEG} were found to increase with the MW, and this effect was higher the greater the difference in molecular size between the two polymers, with a consequent increase of the diagrams' asymmetry.

2.1.2 Temperature

The concentration of phase polymers required for phase separation usually increases with increasing temperature. The experiments of Sjöberg and Karlström [8] suggest that at temperatures below 90 °C, a change in the temperature has only a minor effect on the phase diagram of the PEG/dextran ATPS.

For the PEG/dextran ATPS, the effect of the polymer MWs is further increased with increasing temperature. The STL increased with the temperature due to the fact that $\Delta_{\mathrm{Dextran}}$ decreased with increasing temperature, while Δ_{PEG} remained nearly constant [7].

2.1.3 Inorganic Salts

The hydrophobic (water structure breaking) salts (e.g. $KClO_4$, KI, KSCN) generally elevate the binodal of a two-polymer ATPS, as does the temperature increase, while the hydrophilic (water structure making) salts (e.g. K_2SO_4, KF) depress the binodal of the system [4].

The PEG/dextran system is much less susceptible to the salt effects when compared to polyvinylpyrrolidone (PVP)/dextran or Ficoll/dextran ATPSs. These effects on the phase separation of PEG/dextran seem to be similar to the ones observed on the lower critical solution temperatures (LCSTs) (see Sect. 2.2.1) in the dextran-free aqueous solutions of PEG [4].

The addition of a given salt affects the polymer composition of the two phases depending on the type and total amount of the salt. The salt composition of the phases is also influenced by the total polymer concentration of the system [9]. Bamberger and co-workers [10] found that PEG rejects phosphate, sulphate and to a lesser extent chloride, while the effect of dextran on the distribution of either salt is much smaller. The magnitude of the PEG effect on the salt distribution behaviour was found to be proportional to the polymer concentration.

Zaslavsky and co-workers [9] have established the following empirical relationship between the partition coefficient of the salt (P_{salt}) and the polymer concentration difference of PEG (PVP or Ficoll) in both phases:

$$\ln P_{\text{salt}} = B_{\text{salt}} \Delta C_{\text{PEG}} ,$$

where B_{salt} is a constant depending on the type of the phase polymers and the type and total concentration of salt additive. Hydrophobic salts were found to favour the PEG (PVP or Ficoll)-rich phase, while hydrophilic salts favoured the dextran-rich phase. The STL was related to the total salt concentration in the ATPS [9].

In an ATPS, anions and cations distribute unequally across the interface. To keep the electroneutrality between the phases, a potential difference results [1,5]. Water structure making ions (Li^+, Na^+, NH_4^+, Ca^{2+}, Mg^{2+}, F^-, SO_4^{2-}, CO_3^{2-}, PO_4^{3-}, CH_3COO^-) favour the more hydrophilic phase, whereas water structure breaking ions (K^+, Rb^+, Cs^+, Cl^-, Br^-, I^-, SCN^-, NO_3^-, ClO_4^-) favour the more hydrophobic phase.

In summary, salt additives partition between the two phases and lead to a redistribution of the polymers between the phases, i.e. a change in the phase polymer composition. Therefore, when evaluating the partitioning of proteins in two-polymer ATPSs with salt additives, the interrelationship between the polymer and the ionic composition of the coexisting phases should be taken into account.

2.2 Stimuli-Responsive Soluble–Insoluble Polymers

These systems are based on polymer/water two-phase systems, which are formed due to temperature or pH changes [11,12].

2.2.1 Temperature-Sensitive Polymers

Some polymers exhibit a LCST in water. This means that the solution separates into two phases at temperatures over this point. One of the phases is rich in the polymer and the other phase is poor in the polymer. The LCST is dependent on the polymer concentration and molecular weight.

PEG is a thermoseparating polymer, the LCST of which is around 180 °C for the lower MWs and decreases with increasing MW, reaching a value of approximately 95 °C for MWs of 200 000 or more [13]. Other examples of thermoseparating polymers are the random copolymers of ethylene oxide–propylene oxide, often referred to as EOPO polymers, and poly(*N*-isopropylacrylamide) [poly(NIPAM)], which have much lower LCST values [32 °C for poly(NIPAM)]. The LCST of these types of polymer/water solutions is a linear function of the mass fraction of the PO in the copolymer (Fig. 4) [6]. A PO content of 0% represents the PEG homopolymer.

The LCST diagrams of thermoseparating polymers in aqueous solution have a typical shape, as is represented in Fig. 5. The addition of another component to the polymer/water system may change the clouding behaviour. Hydrophilic additives, such as salts [14–17], glycine [18] and sugars [19], lower the LCST and strongly partition to the water phase after thermoseparation. Cunha and co-workers [20] determined the polymer and salt content in the phases of a 10% EO50PO50 water solution with and without potassium phosphate, after thermoseparation at 50 and 60 °C, respectively. The polymer content in the polymer-rich phase (bottom) reached values between

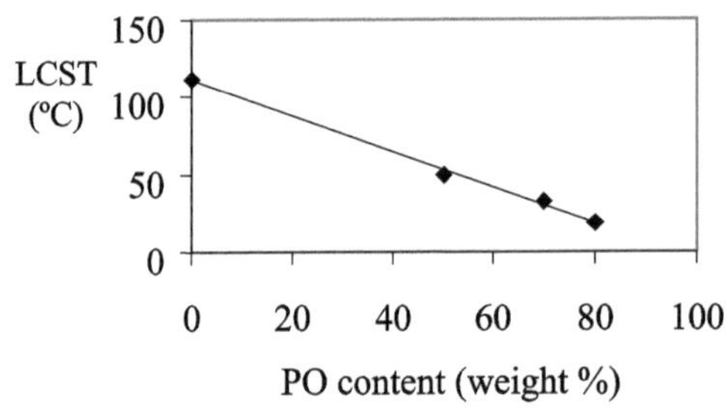

Fig. 4 Lower critical solution temperature of 10% EOPO copolymer/90% water system with the PO content

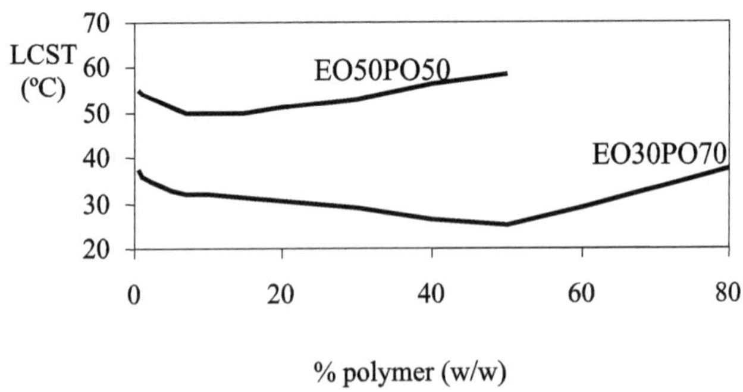

Fig. 5 Lower critical solution temperature diagram for the systems EO50PO50/water and EO30PO70/water

40 and 65% (w/w), and in the water-rich phase, values between 1 and 3% (w/w), depending on the system temperature. Hydrophobic additives, like phenol and *n*-butanol, lower the LCST and partition to the polymer-rich phase [16, 18]. Amphiphilic additives, such as ionic surfactants (e.g. SDS), increase the LCST and nonionic surfactants (e.g. Tween 20) almost do not change the LCST [17].

2.3 Physical Properties of the ATPS

The physical properties of the ATPS (density, viscosity and interfacial tension) determine the phase separation and influence the cell partitioning. Low-density differences between the phases, or highly viscous phases, give rise to high separation times. Systems with a short TLL have low viscosity, but also a low density difference, and therefore they will take a long time to separate. At the other extreme, i.e. large TLL, the density difference is high but the phases' viscosity is also high, leading to long separation times. An intermediate choice will allow a minimum separation time.

Changing the volume ratios, i.e. the dispersed and continuous phases, affects the separation times. When the more viscous phase is dispersed in the less viscous continuous phase (volume of the more viscous phase is the smallest), the separation time is shorter and vice versa.

The interfacial tensions in aqueous polymer systems are very difficult to measure, because they are in the range of 0.5 to 500 $mN\,m^{-1}$, i.e. two to three orders of magnitude lower than those typical for water–organic solvent systems [1]. Low interfacial tensions give rise to small drop diameters, which enable high mass-transfer rates. However, very fine sizes of dispersion lead to long separation times.

3 Factors Affecting Cell Partitioning

Partitioning of cells in ATPSs depends on the interactions of the surface properties of the cells with the physical properties of the two-phase system [1, 21]. The cell surface properties, such as surface charge and lipid composition, and the presence of specific components depend on the cell type. The physical properties of the ATPS, like the interfacial tension, electrostatic potential difference and phase hydrophobicities, however, can be changed by altering the overall composition, namely the type of polymer and ionic composition and concentration [1, 3, 21, 22]. It is difficult to elucidate to what extent each of the physical properties is responsible for the actual cell partitioning by varying the composition of the ATPS, as changing one parameter in the composition of the ATPS can change several of its physical properties.

In the case of PEG/dextran it has been found that by appropriate selection of phase composition, cell separations can be based predominantly on charge-associated or lipid-related properties or on biospecific affinity. The PEG and dextran MWs, the TLL, the type of salts used and the pH are the factors that most likely influence the cell partitioning when no biospecific affinity is used [1, 3, 4].

Cells and cell debris usually partition to the bottom phase, both in PEG/salt and PEG/dextran systems. Cell material contributes to the phase formation, with a consequent decrease of the volume ratio with increasing cell concentration. In PEG/salt systems, nucleic acids and polysaccharides partition strongly into the salt-rich phase. In contrast, most coloured by-products of fermentation broths are hydrophobic and partition into the top phase [23].

The mechanism of partitioning can be described by the Bronsted equation $K = C_{top}/C_{bottom}$, which depends on $e^{\lambda M/kT}$, where M is the molecular weight, k the Boltzmann constant and T the absolute temperature. λ is a factor which depends on properties other than molecular weight (i.e. interactions between the surface properties of the material and the two phases). Albertsson [1] has further proposed that, for particulates, namely cells, the molecular weight should be replaced by the particle's surface area (A) [1, 24]. Thus, size and surface properties determine the partitioning behaviour of biomaterials, these parameters being exponentially related to the partition coefficient.

The partition coefficient of a biomaterial (cell) in an ATPS can be expressed as a function of several variables [1, 6]:

$$\ln K = \ln K^{\circ} + \ln K_{elec} + \ln K_{hfob} + \ln K_{size} + \ln K_{biosp} + \dots,$$

where the subscripts of the several contributions mean: elec, electrochemical; hfob,: hydrophobic; and biosp, biospecific.

These contributions are both from the cell properties (size, net charge, hydrophobicity, other surface properties) and from the surrounding environmental conditions (salts and concentration, pH, type of polymers, polymer molecular weights and concentrations, temperature). K° includes other factors.

3.1 Polymer Molecular Weight and Concentration

The PEG and dextran MWs are expected to have a significant effect on cell partitioning [3] since the MWs of the polymers affect the phase compositions, as previously referred to. However, their effect is small compared to that of the presence of salts, e.g. phosphate. The MW of PEG does not have not a clear effect on hybridoma cell culture in PEG/dextran ATPSs [25]; however, cell morphology was influenced by PEG MW. In cell cultures with PEG 6000, the viable cells were round and similar to the controls, while in cultures with PEG 20 000 and 35 000, the cell shape was more irregular and some very large cells were present.

The influence of dextran MW on cell partitioning is observed in the presence of phosphate. Cell partitioning is a function of both the dextran MW and the TLL at low phosphate concentrations [22]. Cells preferentially accumulate in the lower phase at low TLL in combination with a low dextran MW, which is consistent with the literature data [1, 3].

PEG and dextran have a strong concentration-dependent effect on medium osmotic pressure, while the polymer MW does not have a large influence on osmotic pressure [25]. This is important as high osmolalities decrease cell growth and increase the cell death rate.

The effect of the polymer MW on the partitioning may not be separated from that of the polymer concentration. In addition to Δ_{PEG}, the STL is needed to describe the effect of the polymer MW on partition [4]. At high polymer concentrations (i.e. high interfacial tensions) cells tend to be adsorbed at the interface. As the polymer concentrations are lowered, cells tend to partition between the interface and one of the phases, leading to partition coefficients which depend on the ionic composition and concentration of the phase system and on the surface properties of the cells [21].

3.2 Temperature

The effect of temperature has not been given enough attention and there are very few studies for cells to allow any generalization. The temperature influences the phase composition, the electrostatic and hydrophobic interactions, the macromolecule (protein) conformational state, and it can also induce protein denaturation, self-association or dissociation. The separation of some of these effects is very complex. However for cell processing, the temperature is usually fixed to allow cell culture (37 °C) and/or separation (4 °C to room temperature).

The surface properties of *Escherichia coli* ML308 cells, after heat treatment (from 30 to 42 °C, 0–1 h), were evaluated in PEG/dextran ATPSs. The results suggest that heat treatment induces a change in the cell surface hydrophobicity of *E. coli* cells but not in the cell surface charge [26].

3.3 Salts and pH

Inorganic salts play an important role not only on the physical properties of ATPSs in both polymer–salt and polymer–polymer systems, but also on the cell partitioning.

When phosphate is used, an electrostatic potential difference exists between the PEG and dextran phases with the top PEG-rich phase positive, and cells being pulled out of the interface due to the interaction between the charge of the top phase and the negative surface charge of the cell [21, 27].

On the other hand, when NaCl is used, there is virtually no electrostatic potential difference between the phases and, at higher polymer concentrations, cells remain at the interface. If the interfacial tension of a PEG/dextran ATPS containing NaCl is further reduced (e.g. by reducing the polymer concentrations), cells will partition into the top PEG-rich phase [21]. The partition coefficients obtained in this latter case have been shown to have, at least in the case of red blood cells from different species, an excellent correlation to the ratio of their membrane poly/monounsaturated fatty acids as well as to some other membrane lipid components [28].

The effect of the ratio of potassium phosphate to potassium chloride, defined as the fraction KPi/(KPi + KCl), on cell partitioning also shows a strong indication that the electrostatic potential difference between the two aqueous phases plays a dominant role in the partitioning of hybridoma cells [22]. At high phosphate fractions the potential difference will be relatively high, while at low phosphate fractions it will be almost zero. The electrostatic potential difference is caused by the unequal distribution of buffering ions like mono- and dibasic phosphates between the two phases. These ions partition more to the lower phase and cause a relatively negative potential in the lower phase compared to the upper phase. Chloride ions, which distribute evenly, can reduce the potential difference [1, 3]. The hybridoma cells are "pushed" from the lower phase into the interface with increasing potassium phosphate fractions (increasing the relative negative potential of the lower phase), which was expected, as mammalian cells have a negative overall charge at a pH of around 7.0 [22].

Changing the pH is a common practice in partitioning studies of biomaterials due to the net charge of the biomaterial and its interaction with the surrounding environment. The balance of these interactions may reverse the predictions exclusively based on the biomaterial net charge. Additionally the different pH values are accomplished with different buffers, therefore the salt composition of the ATPS is also altered, and the partitioning of the biomaterials may change accordingly.

Partitioning of cells in PEG/dextran ATPSs is pH- and ionic strength-dependent. However, within the pH range usually used for cell processing (6.6–7.5), it is not expected that the pH would have a large effect on cell partitioning, because the major cell surface charged components usually are fully ionized above pH 5.0 [3, 22]. However, pH strongly influences the ratio of mono- to dibasic phosphates. At pH 7.4 dibasic phosphate is prevalent and causes a higher electrostatic potential difference, leading to partition of the cells into the interface [22]. Also under weak alkaline conditions, when the negative charge on the surface of the cells decreases, the dependence of partitioning on the ionic strength increases significantly [29].

PEG/dextran systems containing various salts have also been used to determine the isoelectric points (pI) of different kinds of cells at different pH values [26]. For example, the dependence of the partition coefficients of *E. coli*

W3110 cells on pH in PEG 4000 (9%)/dextran 100 000–200 000 (9%) containing NaCl, NaSCN or Na_2SO_4, at the same ionic strength of anion, shows that the addition of NaCl and NaSCN led to a negative electrochemical potential, increasing the partition coefficient at low pH values, while the addition of Na_2SO_4 did not. The partitioning of the *E. coli* W3110 cells at different pH values in ATPSs containing different salts of the same ionic strength led to a pI of cells of 2.8 ± 0.3. In general, the surfaces of bacterial and yeast cells have their isoelectric points in the acidic range.

3.4 Affinity Extraction

Aqueous two-phase separation has been used for the fractionation of various biological substances, because of its biocompatibility and easy scale-up features [1]. However, the specificity of the separation is relatively low, which has limited the application of this method to some special cases. Recent developments in biotechnology have raised awareness of the importance of developing specific separation systems especially suitable for use on a large scale and cell and tissue processing for biomedical research and regenerative medicine. In this respect, ATPSs have been modified by introducing affinity ligands to confer suitable specificity [3, 30, 31].

Aqueous two-phase affinity partitioning has been mostly applied to PEG/dextran systems containing PEG derivatives [30, 31]. No affinity two-phase studies have been performed with cells in PEG/salt systems. This is owing to the fact that biospecific interactions are usually obstructed by high salt concentrations. The affinity partitioning is usually performed by addition of the polymer-bound ligand (e.g. PEG-X) to the mixture containing the target biomaterial and the other phase components (e.g. PEG and dextran). After mixing and phase separation, the target biomaterial will be enriched in the phase containing the polymer-bound ligand (top phase). A washing step can follow, by equilibrating the phase containing the ligand and biomaterial (top phase) with the pure phase (bottom). To isolate the target product from the ligand another partitioning step is performed, with a fresh (bottom) phase, under dissociating conditions. Usually pH and salt variations are used, or if a temperature-sensitive polymer is used, by heating above the LCTS of the polymer solution. Cells partition according to the spontaneous pattern, i.e. to the opposite phase, that contains the polymer-bound ligand (bottom).

Several ligands have been developed and applied to the purification of biomaterials. Two major categories of ligands have been distinguished [32]: high affinity (polyclonal and monoclonal antibodies) and general affinity ligands. The latter include ligands such as fatty acids, peptides and triazine dyes which have been used for protein separation; however, they exhibit a wide spectrum of interaction with the target product, and therefore their selectivity is reduced and they can be used in different purification cases. In the case of whole cells, antibodies [29,31], fatty acids [3,24,33], metal ions [30] and dyes [34,35] have been employed as affinity ligands, but in a rather limited manner.

Hydrophobic affinity (fatty acid esters) is mostly used on an analytical scale and for cell separations. The small amounts of modified PEG-fatty acid esters (e.g. the ester of PEG and palmitic acid, PEG-palmitate) usually required have no significant effect on the physical properties of the phases. The different intensities of the hydrophobic interactions between the proteins on the cell surface and the ligand lead to a successful separation of cells to be extracted from the interface into the PEG-rich phase [33].

The immobilized metal ion affinity (IMA) concept described by Porath and co-workers [36] was extended to phase partitioning by grafting iminodiacetate (IDA) onto PEG. This ligand is able to chelate transition metal ions, e.g. Cu(II), Ni(II), Zn(II) and Co(II) [30]. Cells can establish coordination bonds with the immobilized metal ion via accessible and deprotonated histidine residues at the cell surface [30]. In this manner, healthy and pathological cells were separated by IMA partitioning (IMAP) and it was possible to differentiate malarial red blood cells, cancerous fibroblasts and lymphoma cells from their healthy counterparts using PEG-IDA-Ni(II) as an affinity ligand [37]. IMAP systems have been also used to isolate low abundance cells, e.g. human haematopoietic stem cells, from other cell types in a complex cell mixture. Major sources of haematopoietic stem cells are the bone marrow, peripheral blood and cord blood. These cells, representing less than 1% (0.3 to 0.7%) of the total mononuclear cells of human cord blood, are characterized by the expression of the cluster of differentiation (CD) 34 surface antigen. It is known that $CD34^+$ cells are a heterogeneous population, representing the first step of haematopoiesis. They are able to proliferate and to differentiate into the myeloid and lymphoid lineages, and have the capacity for self-renewal in their pluripotent state. The purification of these cells is of great importance for the study of the earliest stages of haematopoiesis and also for the cell therapy of some blood diseases, like leukaemia.

Monoclonal antibodies [38] have been used as ligands to carry out the separation of the complex antigen–antibody and antigen in immunoaffinity partitioning. If antigens or antibodies can be used as a specific ligand, cells can be selectively separated on the basis of cell surface acceptors.

An ATPS has been developed in which a temperature-sensitive polymer [poly(NIPAM)] was used as a ligand carrier for the specific separation of animal cells [29]. Monoclonal antibodies were modified with itaconic anhydride and copolymerized with *N*-isopropylacrylamide, and the ligand-conjugated carriers were added to the PEG 8000/dextran T500 ATPS. The antibody–polymer conjugates were partitioned to the top phase in the absence or presence of 0.15 M NaCl. When ligand-conjugated carriers were used, more than 80% of the cells were specifically partitioned to the top phase in the presence of NaCl up to 0.1 M. The cells were partitioned almost completely to the bottom phase at 0.1 M NaCl or above, when no antibody-conjugate was added in the ATPS. As a model system, CD34$^+$ human acute myeloid leukaemia cells were specifically separated from human T lymphoma cells by applying anti-CD34 conjugated with poly(NIPAM) in the ATPS. By the temperature-induced precipitation of the polymer, about 90% of the antibody–polymer conjugates were recovered from the top phase, which gave approximately 75% cell separation efficiency.

Soluble–insoluble pH-responsive polymers have also been used [38] as functional ligand carriers for the specific separation of several biological substances. In such systems, the ligand-carrier complex is specifically partitioned to the top phases in several ATPSs, and can be easily recovered as a precipitate by changing the pH. ATPSs based on a soluble–insoluble polymer with a ligand carrier to a specific animal cell separation were used for the separation of mouse–mouse hybridoma 16-3F cells, which produce anti-α-amylase antibody [31] from mouse myeloma NS-1 cells. ATPSs composed of 4% PEG 8000 and 5% dextran T500 in 10 mM sodium–potassium phosphate buffer at various pH values and NaCl concentrations were used. To achieve specific separation, Eudragit S 100 and its ligand conjugates were included in the two-phase system. Eudragit is a copolymer of methacrylic acid and methyl methacrylate, with a reversible soluble–insoluble feature dependent on pH. Anti-mouse IgG antibody and thermostable α-amylase as a ligand were coupled to Eudragit using water-soluble carbodiimide. In the ATPS of PEG 8000/dextran T500, the animal cells distributed to the PEG phase under low NaCl concentrations, but partitioned to the bottom phase at NaCl concentrations higher than 0.042 M. Eudragit was distributed to the top phase at NaCl concentrations lower than 0.084 M. With the inclusion of this ligand carrier conjugated with α-amylase, all the cells in the top phase were 16-3F, and no NS-1 cells were detected. On the other hand, the cells in the bottom phase were a mixture of 16-3F and NS-1 cells [31].

4
Applications of the ATPS for Cell Processing

ATPSs have been used for the separation and characterization of different types of cells, namely animal, plant and bacterial cells and viruses. ATPSs are suitable for extractive bioconversions and more recently continuous partitioning of cells has been performed in microdevices, allowing the extension of this methodology to the field of nanotechnology.

One of the emerging areas of application of ATPSs is in animal cell research and cell and tissue engineering for clinical applications, where specific cell separation is a very important technique. For instance, blood contains several species of cells, such as erythrocytes, leukocytes and lymphocytes. In spite of the fact that separation of these cell types is required for basic medical research and therapy, there is no effective method for large-scale cell separation. Furthermore, to establish a hybridoma cell line which has a specific cell surface receptor and produces a specific monoclonal antibody, the concentration and isolation of specific lymphocytes or hybridomas is necessary. Some of these applications are highlighted in the following examples.

4.1 Bacterial Cell Characterization

The membrane surface of bacterial cells is characterized by various properties, such as charge, hydrophobicity and other biospecific properties, due to the lipid–protein composition on the cell membrane. The selection and separation of microorganisms, their cultivation and the removal of cell debris from the target products may all be directly affected by the surface properties of whole cell populations [26]. A study on the extractive cultivation of cells using ATPSs [26, 40–43] showed that quantitative data on surface properties were required to design an efficient process.

The partitioning behaviour of bacterial cells is greatly affected by the addition of salts because of the salting-out effects in ATPSs. While the addition of sodium phosphate increased the partition coefficients, the addition of NaCl led to a decrease of the partition coefficients [43]. With Na_2SO_4, the incremental changes in the partition coefficient of the cell with the addition of salts, $\Delta \ln K_{\text{Cell, Salt}}$, remained almost unchanged. On comparing the effects of the different anions added, phosphate was found to be the most effective in increasing the values of $\Delta \ln K_{\text{Cell, Salt}}$ [43]. The order of effectiveness in increasing the $\Delta \ln K_{\text{Cell, Salt}}$ is then phosphate > SO_4^{2-} > SCN^- > Cl^-. There was also a slight difference in the effect of cations (Na^+ > K^+). This efficiency order corresponds to the Hoffmeister series, which is used to explain the effect of salts on the partitioning of proteins. The partitioning of cells was thus found to depend on the types of salts added as well as the types of bacterial cells used in ATPSs.

4.2 Haematopoietic Cell Separation

Haematopoietic cells, namely erythrocytes, lymphocytes, human haematopoietic stem cells, leukaemia cells and T lymphoma cells, have been selectively separated and characterized by ATPSs.

Erythrocytes from different animals have been partitioned in a PEG/dextran system (with NaCl) containing PEG-palmitate. The partitioning depends on the hydrophobic interaction of the cells with the palmitoyl ligand. Some correlation is also observed between the partition coefficient of the cells and the relative percentages of phosphatidylcholine and sphingomyelin of the cellular membrane, being higher with the former and lower with the latter [3,24,33]. These results reflect not only membrane charge but also membrane lipid composition and affinity for a ligand (i.e. palmitate) [3,24,33].

IMAP was used for the selective separation of human haematopoietic stem cells [30,37]. Partitioning was influenced not only by the relationship between the cell surface and the physical properties of the phases, but also more selectively by the affinity of the immobilized metal ions to the histidine residues of membrane-associated proteins. Nanak and co-workers [37] put forth the hypothesis that glycophorins played an important role in the affinity partitioning of human red blood cells. Nevertheless, on mononuclear cells of cord blood, surface proteins are more diversified and it is more difficult to identify a target molecule for the metal chelate.

4.3 Hybridoma Cell Separation

Hybridoma cell culture and separation have been performed successfully in PEG/dextran ATPSs [22,25], especially by using affinity partitioning [31,34,35].

The partitioning of the mouse–mouse hybridoma cell line BIF6A7, mouse–rat hybridoma PFU-83 and Chinese hamster ovary (CHO) cells DUKX B11-derived cell line BIC-2 in ATPSs of PEG and dextran was studied to identify the key factors governing cell partitioning, and to select ATPSs with suitable cell partitioning for extractive bioconversions with animal cells. The influence of five factors, i.e. the PEG MW, dextran MW, TLL, pH and the fraction KPi/(KPi + KCl), on BIF6A7 cell partitioning was characterized by using a full factorial experimental design [22].

The cell partitioning ranged from complete partitioning into the interface to an almost complete partitioning to the lower phase. In all cases less than 1% of the cells partitioned to the PEG phase. The potassium phosphate fraction had the largest effect on cell partitioning. Low potassium phosphate fractions increased the proportion of cells in the lower phase. To a lesser extent the other factors also played a role in the cell partitioning. The best partitioning for the BIF6A7 cell line was obtained in ATPSs with PEG 35 000, dextran 40 000, TLL 0.10 g/g, pH 7.4 and KPi/(KPi + KCl) 0.1, where 93% of the cells were present in the lower phase [22]. Moreover, the PFU-83 cell line was able to grow in the ATPS hybridoma culture medium. This ATPS hybridoma culture medium therefore seems to be suitable for extractive bioconversions with a wide range of hybridoma cell lines, provided that their product can be partitioned into the upper PEG-rich phase [22].

Zijlstra and co-workers [34,35] also studied the effect ATPS and salt composition on the affinity partitioning of mouse–mouse hybridoma BIF6A7 cells and its monoclonal antibody IgG class 2a in PEG 35 000/dextran 40 000 containing a PEG–dye (mimetic green 1 A6XL) ligand. In the presence of 1% PEG–dye ligand, the binding of IgG was affected by sodium chloride concentration. When no PEG ligand was present, the cells partitioned preferentially to the bottom phase, while in the presence of the PEG ligand they partitioned almost completely into the interface.

4.4 Extractive Bioconversions in Aqueous Two-Phase Systems

Extractive bioconversions [44] have been successfully applied to integrate the production step (fermentation, cell culture, bioconversion media) with the first steps of downstream processing of a target product, by increasing the overall productivity and product yield. ATPSs are suitable techniques for in situ product recovery due to their mild characteristics and biocompatibility. For efficient processing the cell (biocatalyst) has to partition completely to one of the phases, while the product concentrates in the other phase. In this manner the biocatalyst can be retained and reused by recycling the biocatalyst/cell-containing phase, and a cell-free product phase is available for further downstream processing.

Several bacterial species have been cultured in polymer–salt ATPSs [40–43, 45]. Polymer–salt ATPSs can be cost effective for protein extractions [45, 46]. Furthermore, their suitability for large-scale extractions of products with a wide range of hydrophobicities has long been recognized [46]. Cell growth was found to be largely unaffected by PEG when the molecular weight was 4000 kD or higher; however, it was inhibited exponentially with increasing salt concentration [45].

For extractive bioconversions with animal cells excreting high-value proteins, such as monoclonal antibodies (MAbs) [25, 47], the presence of cells in the interface is not desirable. Therefore, for extractive bioconversions ATPS culture media should be designed in which the cells partition to the lower dextran-rich phase. In ATPSs of PEG 35 000 and dextran 40 000 with culture medium, long-term growth of hybridoma BIF6A7 cells has been shown with the cells partitioned almost completely to the lower phase [25]. The MAb product (IgG class 2a), however, partitioned along with the cells in the lower phase. An affinity dye ligand (mimetic green 1 A6XL) coupled to PEG led to an improvement of the partition coefficient of the MAb to the top phase [33, 34].

Plant cell cultivation in ATPSs has also been reported in ATPSs with neutral polymers [48, 49]; however, product excretion, required for successful in situ product removal, remains an issue to be improved.

4.5 Continuous Cell Partitioning in Microfluidic Devices

Recently, microfluidic devices, in which an aqueous two-phase laminar flow is stably formed, have been used for the continuous partitioning and fractionation of animal and plant cells [50, 51]. This method utilizes the stable aqueous two-phase laminar flows and high surface-to-volume ratio within a microchannel, without diffusive mixing and making the interface large enough to perform continuous partitioning.

Live and dead CHO-K1 cells were fractionated by continuous-flow extraction in a microfluidic device using ATPSs of PEG 8000 (4%) and dextran T500 (5%) at various pH values [50]. The fractionation efficiency of live and dead CHO-K1 cells from the culture broth was compared in a normal macroscale system and microfluidic device. The optimum pH for the fractionation was 6.6 in both the normal and microscale systems. The recovery efficiency of live cells was 85.5% in the normal macroscale ATPS, while all of the live CHO-K1 cells were recovered to the PEG-phase flow in the microfluidic device. The fractionation efficiency of live cells was 97.0%. Only 3% of the dead cells were recovered from the cell suspension to the PEG-rich phase. The continuous-flow two-phase system showed better recovery and fractionation of cells than the normal macroscale two-phase system.

The continuous two-phase partitioning of cultured plant cell aggregates with diameters between 37 and 96 μm was also performed in microfluidic devices [51]. The cell partitioning was performed in a simple microchannel having two inlets and two outlets, and the effects of the widths of each phase and the flow rate on the partitioning efficiency were examined. In addition, for the improvement of the partitioning efficiency, partitioning in a microchannel with a pinched segment was also performed, suggesting that this microscale aqueous two-phase flow system can further be incorporated into micro total analysis systems or lab-on-a-chip technology, owing to its simplicity, applicability and biocompatibility.

This partitioning method seems to be highly advantageous for the separation and selection of cell lines or biomaterials from heterogeneous mixtures, such as stem cells or highly producing plant cells, since partitioning can be continuously performed in a relatively short time under mild conditions. In addition, by combining with other miniaturized manipulation systems for sampling, analysis or cell cultivation, this method will pave a new way for the development of biotechnology and tissue engineering, thus eliminating time- and labour-consuming procedures.

References

1. Albertsson P (1986) Partitioning of cell particles and macromolecules, 3rd edn. Wiley, New York
2. Kula MR (1985) Liquid–liquid extraction of biopolymers. In: Cooney CL, Humphrey AE (eds) Comprehensive biotechnology. The principles of biotechnology: engineering considerations, vol 28. Pergamon, New York, p 451
3. Walter H, Brooks DE, Fischer D (eds) (1985) Partitioning in aqueous two-phase systems. Academic, Orlando, FL

4. Zaslavsky BY (1995) Aqueous two-phase partitioning: physical chemistry and bioanalytical applications. Dekker, New York.
5. Diamond AD, Hsu JT (1992) Aqueous two-phase systems for biomolecule separation. In: Fiechter A (ed) Advances in biochemical engineering/biotechnology. Springer, Heidelberg, p 89
6. Cunha MT, Aires-Barros MR, Cabral JMS (2003) Extraction for rapid protein isolation. In: Kaul R, Mattiasson B (eds) Isolation and purification of proteins. Dekker, New York, p 321
7. Forciniti D, Hall CK, Kula MR (1991) Fluid Phase Equilib 61:243
8. Sjöberg A, Karlström G (1989) Macromolecules 22:1325
9. Zaslavsky BY, Miheeva LM, Aleschko-Ozhevski YP, Mahmudov AU, Bagirov TO, Garaev ES (1988) J Chromatogr 439:267
10. Bamberger S, Seaman G, Brown JA, Brooks DE (1984) J Colloid Interface Sci 99:187
11. Galaev IY, Mattiasson B (1993) Enzyme Microb Technol 15:354
12. Mattiasson B, Galaev IY (1999) Bioseparation 7:175
13. Bailey FE, Koleske JV (1976) Poly(ethylene oxide). Academic, New York
14. Florin E, Kjellander R, Eriksson JC (1983) J Chem Soc Faraday Trans 80:2889
15. Ananthapadmanabhan KP, Goddard ED (1986) J Colloid Interface Sci 113:294
16. Louai A, Sarazin D, Pollet G, François J, Moreaux F (1991) Polymer 32:713
17. Cunha MT, Tjerneld F, Cabral JMS, Aires-Barros MR (1998) J Chromatogr B 711:53
18. Johansson H-O, Karlström G, Tjerneld F (1997) Colloid Polym Sci 257:458
19. Sjöberg A, Karlström G, Tjerneld F (1989) J Am Chem Soc 22:4512
20. Cunha MT, Cabral JMS, Aires-Barros MR (1997) Biotechnol Tech 11:351
21. Walter H, Anderson JL (1981) FEBS Lett 131:73
22. Zijlstra GM, Michielsen MJ, de Gooijer CD, van der Pol LA, Tramper J (1996) Biotechnol Prog 12:363
23. Hustedt H, Kroner KH, Menge U, Kula MR (1985) Trends Biotechnol 3:139
24. Walter H, Fischer D, Tilcock C (1990) FEBS Lett 270:1
25. Zijlstra GM, de Gooijer CD, van der Pol LA, Tramper J (1996) Enzyme Microb Technol 19:2
26. Umakoshi H, Kuboi R, Komasawa I (1997) J Ferment Bioeng 84:572
27. Zaslavsky BY, Miheeva LM, Mestechkina NM, Pogorelov VM, Rogozhin SV (1978) FEBS Lett 94:77
28. Walter H, Krob EJ, Brooks DE (1976) Biochemistry 15:2959
29. Kumar A, Kamihira M, Galaev IY, Mattiasson B, Iijima S (2001) Biotechnol Bioeng 75:570
30. Laboreau E, Capiod JC, Dessaint C, Prin L, Vijayalakshmi MA (1996) J Chromatogr B 680:189
31. Hamamoto R, Kamihira M, Iljima S (1996) J Ferment Bioeng 82:73
32. Labrou N, Clonis YD (1994) J Biotechnol 36:95
33. Walter H, Krob EJ (1977) FEBS Lett 78:105
34. Zijlstra GM, Michielsen MJM, de Gooijer CD, van der Pol LA, Tramper J (1998) Bioseparation 7:117
35. Zijlstra GM, Michielsen MJM, de Gooijer CD, van der Pol LA, Tramper J (1996) Bioseparation 6:201
36. Porath J, Carlsson J, Olsson I, Belfrage G (1975) Nature 258:598
37. Nanak E, Vijayalakshmi MA, Chadha KC (1995) J Mol Recognit 8:77
38. Elling L, Kula MR, Hadas E, Katchalski-Katzir E (1991) Anal Biochem 192:74
39. Kamihira M, Kaul R, Mattiasson B (1992) Biotechnol Bioeng 40:1381
40. Kuboi R, Umakoshi H, Komasawa I (1995) Biotechnol Prog 11:202

41. Persson L, Staaibrand H, Tjerneld F, Hahn-Hagerdal B (1991) Appl Biochem Biotechnol 27:27
42. Tanaka H, Kuboi R, Komasawa I, Tsuchido T (1993) J Ferment Biotechnol 75:424
43. Umakoshi H, Yano K, Kuboi R, Komasawa I (1996) Biotechnol Prog 12:51
44. Kaul R, Mattiasson B (1991) Bioproc Technol 11:173
45. Kuboi R, Maruki T, Tanaka H, Komasawa I (1994) J Ferment Bioeng 78:431
46. Kula MR (1990) Bioseparation 1:181
47. Zijlstra GM, Michielsen MJM, de Gooijer CD, van der Pol LA, Tramper J (1998) Curr Opin Biotechnol 9:171
48. Buitelaar RM, Leenen EJTM, Tramper J (1992) Biocatalysis 6:73
49. Ilieva MP, Bakalova A, Mihneva M, Pavlov A, Dolapchiev L (1996) Biotechnol Bioeng 51:488
50. Nam KH, Chang WJ, Hong H, Lim SM, Kim DI, Koo YM (2005) Biomed Microdevices 7:189
51. Yamada M, Kasim V, Nakashima M, Edahiro J, Seki M (2004) Biotechnol Bioeng 88:489

Adv Biochem Engin/Biotechnol (2007) 106: 173–193
DOI 10.1007/10_2006_043

Published online: 13 January 2007

Development of Separation Technique for Stem Cells

Masamichi Kamihira[1] (✉) · Ashok Kumar[2,3]

[1]Department of Chemical Engineering, Faculty of Engineering, Kyushu University, 744 Motooka, Nishi-ku, 819-0395 Fukuoka, Japan
kamihira@chem-eng.kyushu-u.ac.jp

[2]Department of Biological Sciences and Bioengineering, Indian Institute of Technology, 208016 Kanpur, India

[3]Department of Biotechnology, Center for Chemistry and Chemical Engineering, Lund University, S-221 00 Lund, Sweden

Abstract In recent years, human embryonic stem cells have been established, and somatic stem cells derived from various adult organs have been identified and characterized to differentiate into various kinds of functional cells. There have been attempts to use functional cells induced from such stem cells for tissue regeneration and cell therapy. The method is expected to become an important treatment for intractable diseases in the near future. Since tissues and organs generally contain only a small quantity of somatic stem cells, and since it is necessary to separate functional cells generated from stem cells for use in therapy, an effective method for specific cell separation is crucial to the practical application of regenerative medicine. For the specific separation of cells, a fluorescence activated cell sorter using specific antibodies is a powerful tool, but the method is not suitable for large-scale processing and a special device is required. Although a magnetic cell separation system using immuno-magnetic fine particles is also commercially available, the system still needs special apparatus for large-scale processing. We developed a novel method for the separation of specific cells in an aqueous two-phase system using antibodies modified with a temperature-responsive polymer. The method enables the processing of a large quantity of cells without the requirement of a special device.

Keywords Aqueous two-phase system · Immunoaffinity partitioning · Poly-*N*-isopropylacrylamide · Specific cell separation · Stem cell

1 Introduction

In recent years, stem cells such as neural stem cells and mesenchymal stem cells have been identified in various organs. Many studies on the proliferation and differentiation of stem cells in vitro have been carried out in preparation for therapy by cell transplantation. In addition, now that human embryonic stem (ES) cells have been established [1, 2], it is expected that all kinds of tissues and organs will be reproduced using ES cells [3–5]. It seems that advances in stem cell technology will enable the restoration of tissues and organs using constructed cell transplantation as a therapy for intractable diseases in the near future [3, 4].

The process of cell separation holds an important role in cell therapy and regenerative medicine using stem cells. Stem cells are usually present in only small quantities in adult tissues and organs, and an effective separation procedure for stem cells is always required. Furthermore, even if one succeeds in the separation and/or establishment of stem cells and ES cells and subsequently expansion and differentiation into functional cells by setting suitable culture conditions, specific cell separation is still required, since only differentiated functional cells should be used for practical treatments.

The separation of cells is different from the separation of other materials. Proteins, for example, are relatively unstable and must be carefully handled not to lose their biological activity during the separation process, but proteins can be purified as a homogeneous material in terms of molecular weight and structure. On the other hand, in cell separation, careful handling is also important. In addition, since living cells may change during the separation process, it is difficult to separate cells in completely the same state. Instead, the purpose of cell separation is to collect a group of cells sharing the same characteristics. For example, in the case of the separation of T lymphocytes in blood, the cells are collected from various blood-constitutive cells in accordance with some definition of T lymphocytes. Such cells are not 100% pure, but rather belong to a group sharing a characteristic of T lymphocytes, and may be further classified into subgroups. Since cells are living and variable depending on the surrounding environment, the separation processes are limited by the medium and operation. Rapidness of the operation and viability during separation are important operating factors. Thus, for purity and recovery in the separation of specific cells, it is necessary to wrestle with problems different from those in the separation of a material such as protein.

In this chapter, after summarizing recent progress in stem cell technology for regenerative medicine and separation techniques for specific cells, we describe a separation method for stem cells and specific cells using aqueous two-phase systems and specific antibodies modified with a synthetic polymer.

2
Stem Cells and Regenerative Medicine

Advances in medicine and medical technology have resulted in a tremendous improvement in health and welfare. However, we are still faced with various diseases that are difficult to treat using contemporary medicine. For organ failures (heart failure, renal failure, liver failure) and neurodegenerative diseases (Parkinson's and Alzheimer's disease), there is at present no effective treatment other than the transplantation of organs from human donors or cells from a fetus. In the case of transplantation, there are many problems such as immunological rejection, infectious diseases, and a lack of donors, and the development of a novel treatment method has been desired. During the past decade, regenerative medicine has appeared as a key technology for the next generation of medical care [6–12]. Cell therapy and organ repair using stem cells have become very attractive in regenerative medicine.

A stem cell is generally defined as a cell which can renew and maintain itself for a long period of time with the potential to commit to a cell or tissue lineage with specialized functions (Fig. 1). A fertilized egg and embryonic cells in the early stages (to morula) have the ability to form all cell types including the entire fetus and placenta. Thus, such cells can basically form whole organs (totipotent). On the other hand, stem cells whose ability to differentiate is limited to some extent can be acquired from a developing embryo. Such cells have the potential to differentiate into all cell types composing the human body (pluripotent) in addition to the self-renewal capacity. ES cells and embryonic germ (EG) cells are classified in this category, and have the ability to differentiate in vitro into cells of all somatic cell lineages as well as into germ cells. In recent years, it has been discovered that various multipotent stem and progenitor cells exist in adult tissues and organs to replace lost or injured cells [6, 10–17]. Thus, there has been tremendous progress in the understanding of the mechanism of tissue regeneration. Representative somatic stem cells that have been identified and characterized are neural stem cells in the brain, hematopoietic and mesenchymal stem cells in bone marrow, and stem cells in each organ (e.g., hepatic oval cells, pancreatic stem cells) (Fig. 1). If it becomes possible to isolate stem cells and to control their differentiation, theoretically, all tissues and organs will be able to be regenerated from ES cells, and a limited variety of tissues and organs will be able to be regenerated from somatic stem cells (Fig. 2). Human ES cell lines have been established, and are expected to serve as an unlimited source of cells for regenerative medicine. Somatic stem cells in adult tissues seem to play a role in the maintenance of homeostasis. One example is hematopoietic

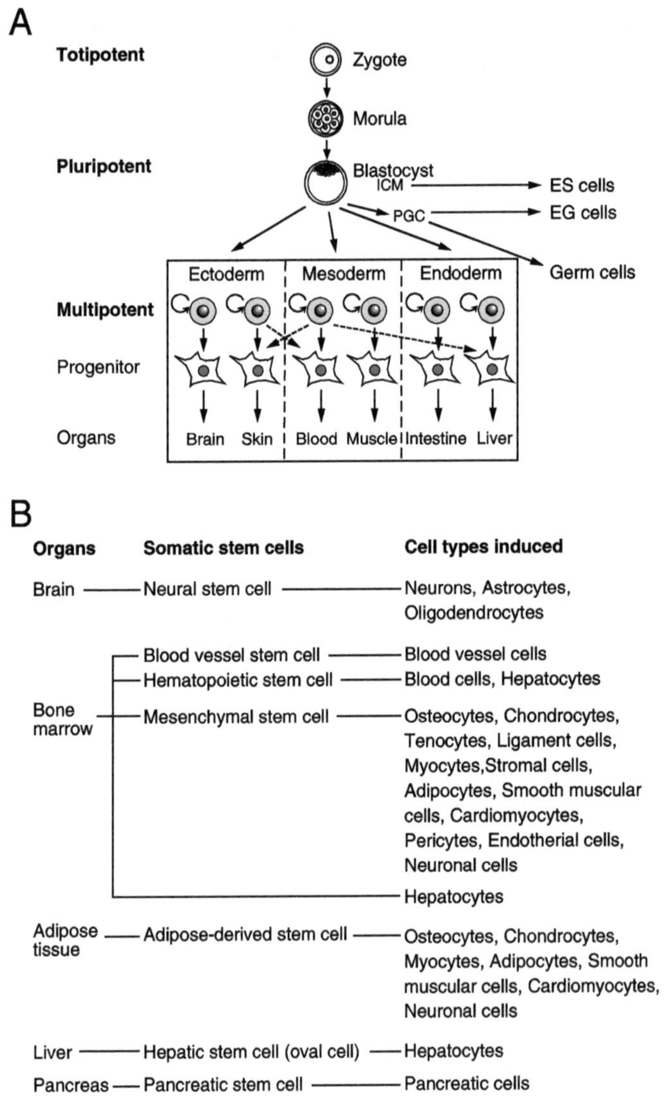

Fig. 1 Classification of stem cells. **A** Hierarchy of stem cells. **B** Examples of somatic stem cells

stem cells, which continue to supply blood cells through self-renewal and differentiation. The transplantation of hematopoietic stem cells has been used for tissue regeneration in the clinical setting.

Recent studies have revealed that multipotent somatic stem cells from adult tissues and organs are almost comparable to ES cells with respect to their ability to differentiate into various tissues in vitro and in vivo [9, 18–20]. The function is termed "stem cell plasticity". For example, bone marrow-

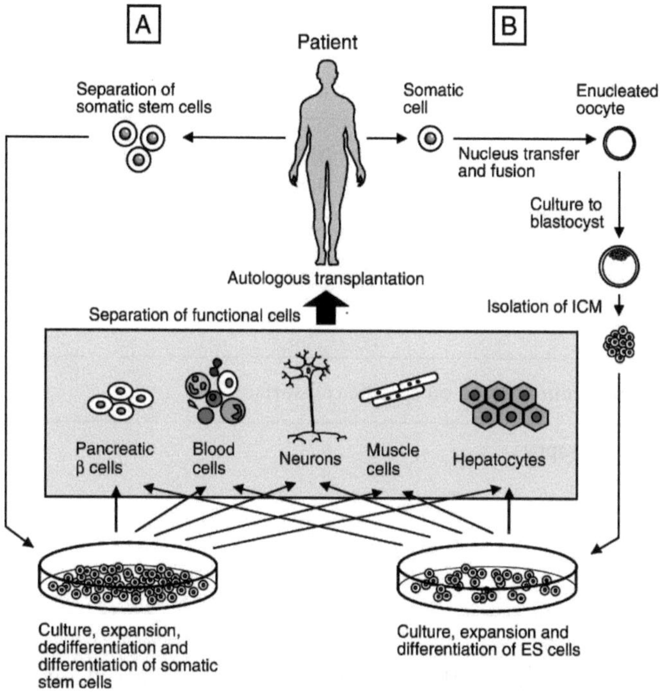

Fig. 2 Strategy of autologous tissue and organ regeneration and cell therapy using somatic stem cells (**A**) and ES cells (**B**)

derived mesenchymal stem cells can differentiate not only into osteoblasts, chondrocytes, myocytes, tenocytes and adipocytes, but also into neuron-like cells. Some sorts of stem cells such as neural stem cells isolated from brain can be cultured to multiply now. Further comprehensive studies on the growth and differentiation of somatic stem cells will make it possible to use them for biomedical research, tissue engineering, regenerative medicine, and organ repair in the near future.

3
Procedures for Cell Separation

Various procedures have been used to separate cells [21–23]. The procedures applied to the separation of substances such as proteins have been used for cell separation as well. Cell sorting (flow sorting) based on flow cytometry was developed as a specialized technique for specific cell separation using an automated device. Methods used for cell separation are summarized in Table 1. They are classified into two categories: those methods based on a difference in a physicochemical property such as density and electrostatic and

Table 1 Cell separation procedures

A. Separation procedures based on physicochemical properties	
Centrifugation	Differential centrifugation Density gradient centrifugation
Unit gravity sedimentation	
Precipitation	
Aqueous two-phase extraction	
Filtration (glass wool, nylon wool)	
Electrophoresis	
B. Separation procedures based on specific cell-surface	
Affinity chromatography	
Affinity filtration	
Affinity partitioning	
Affinity precipitation	
Flow sorting	
Magnetic separation	

hydrophobic character of the cell surface and those based on the expression of a marker protein. Centrifugation, unit gravity sedimentation, aqueous two-phase extraction, and electrophoresis are classified in the former category, and affinity chromatography, affinity filtration, affinity partitioning, flow sorting, and magnetic separation using ligand-modified magnetic particles are classified in the latter.

The procedures often used for separation of stem cells – density gradient centrifugation, flow sorting, and magnetic separation – are described in detail.

3.1 Density Gradient Centrifugation

Centrifugation is frequently employed for bulk treatment of cells rather than precise separation [21, 24, 25]. Density gradient centrifugation is superior to differential centrifugation, which utilizes a difference in sedimentation speed based on the density of cells (or size), in terms of separation performance and cell recovery. While differential centrifugation is performed in a homogeneous solution, density gradient centrifugation is conducted in a solution with a heterogeneous density gradient to make bands of cells in the solution corresponding to the density and size of cells. There are two modes of operation in density gradient centrifugation; rate-zonal and isopycnic (equilibrium) (Fig. 3). In rate-zonal centrifugation, the cell suspension is placed on the top of the resolution solution forming a density gradient. Under centrifugal force, cells move in the solution with a different speed depending on

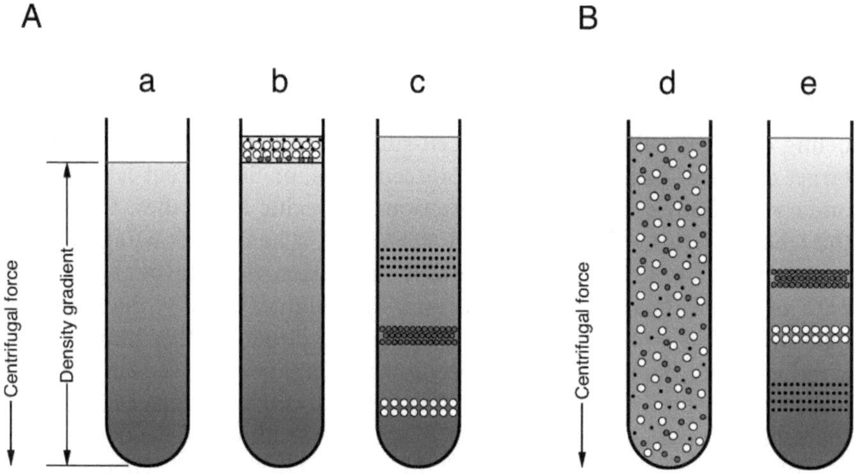

Fig. 3 Operation modes in density gradient centrifugation. **A** Rate-zonal centrifugation. (*a*) A centrifuge tube is filled with a density gradient solution. (*b*) A cell suspension is added to the top of the gradient solution. (*c*) Cells move with a different speed depending on cell size during centrifugation. **B** Isopycnic centrifugation. (*d*) A cell suspension is mixed with a gradient material. Alternatively, cells are applied to the top of a density gradient solution. (*e*) After centrifugation, a density gradient forms in the solution, and cell bands are formed at a density balanced with the density of cells

their size, and separate from each other to make bands consisting of a cell group sharing a similar sedimentation speed (size). The density gradient of the solution increases at the bottom and the density of the solution is set up below the density of cells. Therefore, when centrifugation continues, all cells are pelleted at the bottom. On the other hand, with isopycnic centrifugation, the density of the solution at the bottom is greater than the density of the cells. Consequently, cells do not reach the bottom even after centrifugation for a long time, and bands of cells are formed at a density balanced in the gradient to the density of cells. Cells are applied to the top of the density gradient solution and then centrifuged for separation. Alternatively, the solution containing a gradient material and cells are mixed first, and then centrifugal force is applied to make a density gradient during cell separation. However, the latter procedure is rarely performed, since the processing takes a long time. Since type-specific separation is difficult with this method, it is often used for pretreatment in the separation of stem cells.

3.2
Flow Sorting Using a Flow cytometer

Flow sorting is a specialized cell separation procedure in which cells are analyzed and sorted at the same time using an automated cell analysis technology based on flow cytometry [26–32]. It is a well-known method for precise cell sorting, generally using an automatic device called a fluorescence activated cell sorter (FACS). Target cells are labeled with fluorescence, and the device separates cells according to the strength of fluorescence. For the labeling of target cells, antibodies to cell-surface marker proteins expressed in target cells are conjugated with fluorescent dyes, or fluorescent-labeled secondary antibodies are used for detection of specific antibodies. Cells after the labeling treatment are subjected to FACS. A schematic drawing of FACS is shown in Fig. 4. Each cell in the cell suspension flows through the light path of a laser beam. During this time, light signals such as fluorescence, light-scattering and absorbance are measured. Based on this information, a cell emitting fluorescence at a specific wavelength with a certain strength is selected instantly, and the liquid flow is cut to form a droplet containing the cell. It is possible to collect the droplet by sending an electric current to polarized boards, since a liquid drop is electrostatically charged. The

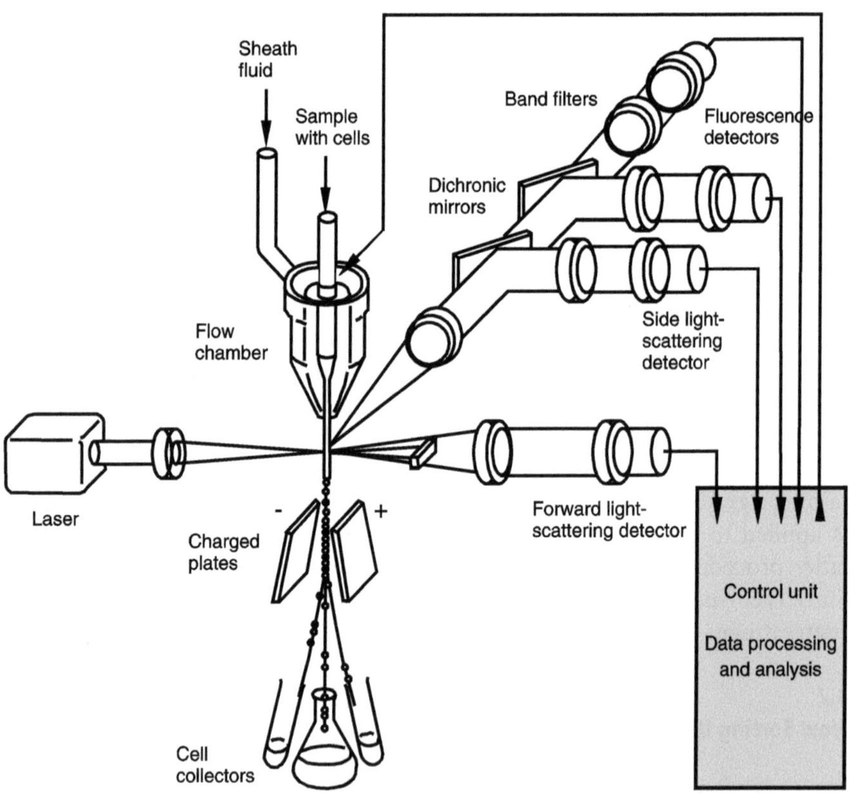

Fig. 4 Schematic drawing of the flow cytometer

Table 2 Identification markers for human stem cells

Stem cells	Markers
Hematopoietic stem cells	Lin(–), CD34, CD133, CDCP1, C-KIT, KDR, VEGFR1
Mesenchymal stem cells	CD49a, CD45(med/low), CD166, CD105, CD73, HLA-ABC, HLA-DR(–)
Adipose-derived stem cells	CD9, CD29, CD49d, ICAM-1(CD54), CD105, ALCAM(CD166), CD44, CD71, CD90, CD146, CD55, CD59, HLA-ABC, HLA-DR(–)
Neural stem cells	GFAP*, Sox1*, Sox2*, Nestin*, Musashi*
ES cells	SSEA-3/-4, TRA-1-60/81, GCTM-2, TG-343, TG-30, CD9, CD133, ALP*, Oct-4*, Nanog*, Sox2*

* Intracellular markers

fractionation of cells is achieved one cell at a time. The advantage of this method is high sensitivity and high resolution, since an individual cell is processed based on the fluorescence of each cell using a highly specific antibody. By labeling cells with several kinds of antibodies and fluorescent dyes at the same time, further complicated separation is also possible. Marker proteins for stem cells are summarized in Table 2. For example, CD34 is the most commonly used marker to obtain enriched population of human hematopoietic stem cells and progenitors. In the case where specific cell-surface markers to sort cells are not available, a reporter construct, in which a fluorescent protein expressed under the control of a specific promoter is inserted, is introduced into cells to make a fluorescent mark on specific cells [7].

3.3 Magnetic Separation

In the magnetic separation of cells, specific cells are first labeled with fine magnetic particles, and then the labeled cells are separated using a magnet [17,33–39]. The procedure is theoretically very simple, and is very effective if successful. The labeling of target cells with fine magnetic particles is accomplished using specific antibodies modified with magnetic particles. The particles should be small enough not to self-aggregate, but large enough for easy magnetic separation when they adsorbed on cells. In addition, the particles should exhibit minimal nonspecific adsorption to cells and ease of conjugation to antibodies. Magnetic particles 0.2–2 μm in diameter are frequently employed. The separation of cells labeled with magnetic particles is achieved using either a solid magnet or an electromagnet. A device using an electromagnet is presented in Fig. 5 [33]. A magnetic field produced by an electromagnet is applied to a glass column filled loosely with fine stain-

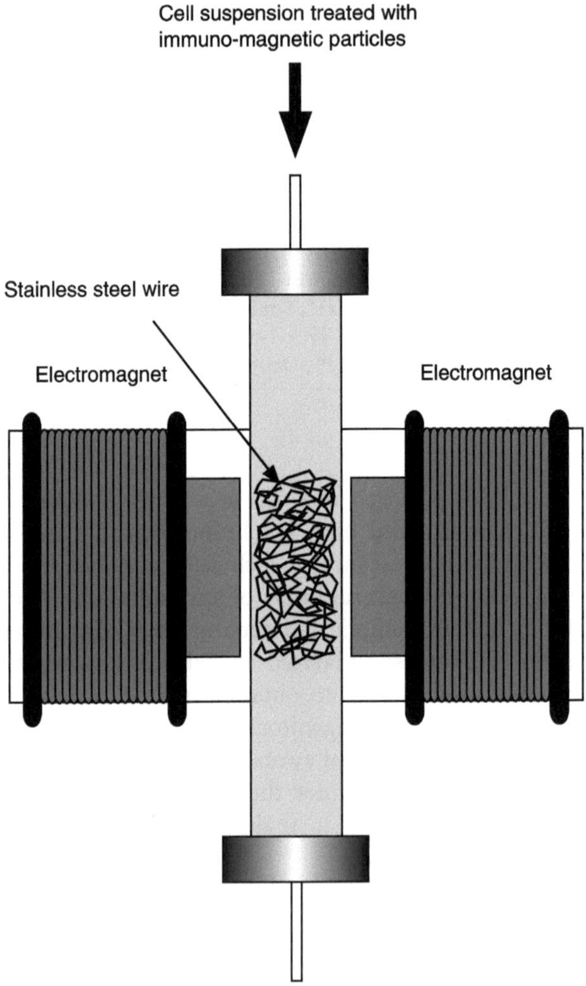

Fig. 5 Apparatus for the magnetic separation of cells

less steel wire. When the target cells labeled with magnetic particles pass through the column, they are attracted to the surface of the steel wire. The magnetic field formed when the steel wire is magnetized declines drastically away from the surface of the steel wire. Thus, a large magnetic field gradient is formed in the space between wires. This magnetic field gradient is not generated without a steel wire, even using a powerful magnet. The steel wire should be loosely filled so that cells are not captured mechanically, but the wire bed should be long enough for magnetic capture of cells because the distance of the magnetic field forming between wires is short.

4
Specific Cell Separation Using Affinity Partitioning in Aqueous Two-Phase Systems

As described above, in order to separate specific cells as a group sharing a specific function, the cells marked using a fluorescent-labeled antibody against a cell surface protein are isolated using a specific apparatus such as a flow cytometer. With this method, several kinds of specific antibodies can be used for the labeling of cells at the same time, and thus a precise separation of cells based on the kind and amount of marker proteins expressed on the cell-surface can be performed. However, the equipment needed is expensive and so the method is unsuitable for mass processing on a large scale. For cell separation in the field of regenerative medicine, a precise fractionation achieved only by flow cytometry is not required in most cases, but it is assumed that large quantities and quick processing will be demanded. Therefore, simple procedures such as centrifugation and sedimentation are still the mainstream of cell separation in this field. However, such procedures are insufficient for the separation of a cell group sharing some function, and there are many needs for the specific separation of functional cells based on the sorting ability of an antibody. In recent years, a cell separation system based on magnetism using antibody-modified fine magnetic particles has attracted a great deal of attention in the field of cell separation, but a special device for the magnetic separation of cells is necessary for large scale processing.

As a simple method without needing a special device, we developed a cell separation procedure using aqueous two-phase systems and an antibody modified with a synthetic polymer for the specific partitioning of target cells [40].

4.1
Aqueous Two-Phase System

Aqueous two-phase systems, in which both phases contain more than 80% water, are formed when a water-soluble polymer such as polyethylene glycol (PEG) is mixed with dextran or salts in water. Liquid-liquid extraction in the aqueous two-phase system has provided a gentle procedure for the separation of biological substances such as proteins, organelles and cells, without the loss of their biological activity [41–43]. Although we cannot apply the system to cell separation if the components seriously affect cell viability, aqueous two-phase systems comprising PEG-dextran have low toxicity for cells in practical concentrations [44–48]. In addition, the separation method is scalable and suitable for large-scale separation from a characteristic of liquid-liquid extraction. To date, various biological substances have been used for separation with aqueous two-phase systems, and specific partitioning of the target to one of the phases has been a major problem. In order to achieve specific partitioning, an affinity ligand, which is unmodified or modified to partition to

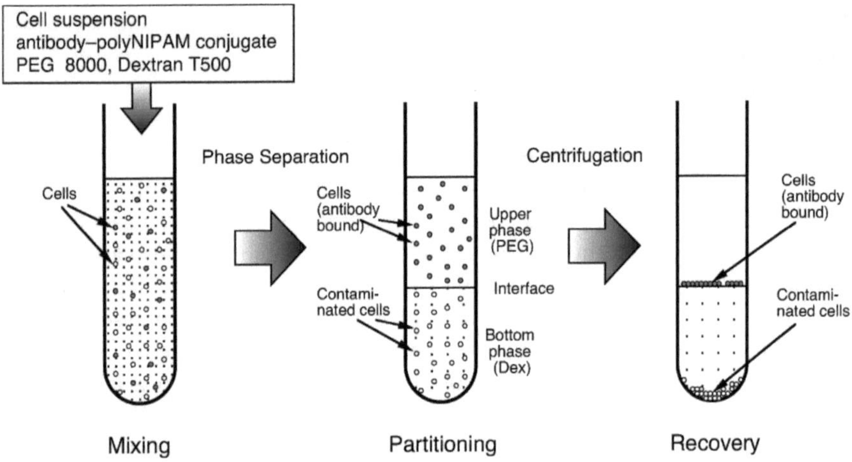

Fig. 6 Scheme for cell separation using an aqueous two-phase system and thermo-responsive polymer-antibody conjugate

one of the phases (mostly the upper phase), has been introduced into the two-phase system. In the separation system, a target substance bound to an affinity ligand is specifically partitioned to the phase to which the affinity ligand is partitioned. As the ligand, fatty acids [49], peptides [50], triazine dyes [51, 52], and metal ions [53–55] have been used for protein separation. Protein ligands such as antibodies have been also used for a highly specific separation [45, 56, 57]. Protein ligands are often modified with an upper phase polymer, PEG, pH-sensitive or temperature-sensitive soluble-insoluble polymers, or magnetic particles to promote the specific partitioning and easy separation [58–61].

In our separation system, a monoclonal antibody against a cell surface protein was modified with a temperature responsive polymer, poly-*N*-isopropyl acrylamide (polyNIPAM) [62, 63], because of the low non-specific adsorption character and sharp temperature response behavior [64]. Using the polyNIPAM-modified antibody partitioned to the upper-phase in the PEG-dextran aqueous two-phase system, we partitioned cells expressing the surface antigen to the upper phase, and specific cells could be concentrated and separated from other cells at the interface of the two phases after applying gentle centrifugation (Fig. 6).

4.2 Modification of Antibody with PolyNIPAM

An outline of the modification procedure for antibodies with polyNIPAM is shown in Fig. 7. First, antibodies dissolved in phosphate buffer were treated with itaconic anhydride in the presence of glucose to introduce C – C dou-

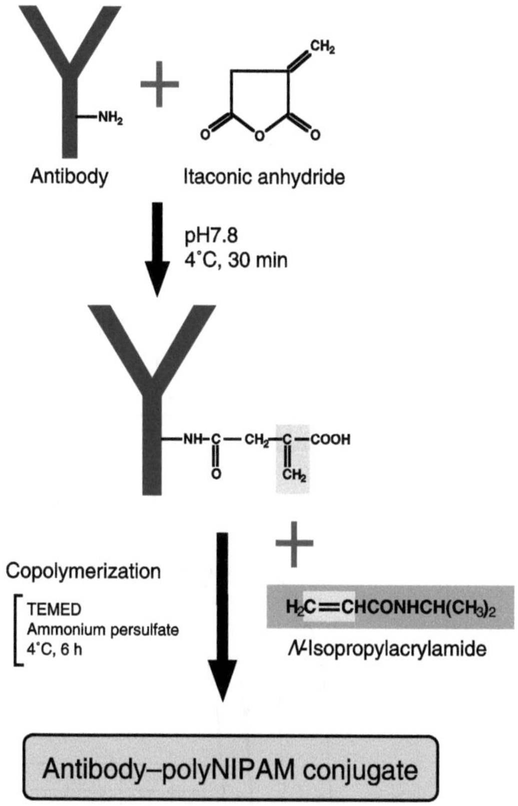

Fig. 7 Modification of antibody with polyNIPAM

ble bonds into multiple amino groups within an antibody. Successively, the derivatized antibodies were copolymerized with NIPAM by radical copolymerization to produce antibodies modified with many polyNIPAM chains mediated by itaconate. With the optimization of the amount of NIPAM during the reaction, polyNIPAM-modified antibodies were obtained with 80% yield without the loss of antibody activity. PolyNIPAM is temperature-sensitive, and its soluble-insoluble character in water can be controlled by a change in temperature. Figure 8 shows temperature responsive behavior of native polyNIPAM and anti-CD34 antibodies modified with polyNIPAM. PolyNIPAM precipitates from aqueous solutions at 32–35 °C as a result of the progressive increase in intra- and intermolecular hydrophobic interactions with elevating temperature. This temperature is called the cloud-point or lower critical solution temperature (LCST) of the polymer. The homopolymer of polyNIPAM showed a sharp phase transition at around 33 °C. The conjugation of the polymer with protein molecules makes it slightly hydrophilic compared with native polyNIPAM, and thus elevated the LCST of the copolymer.

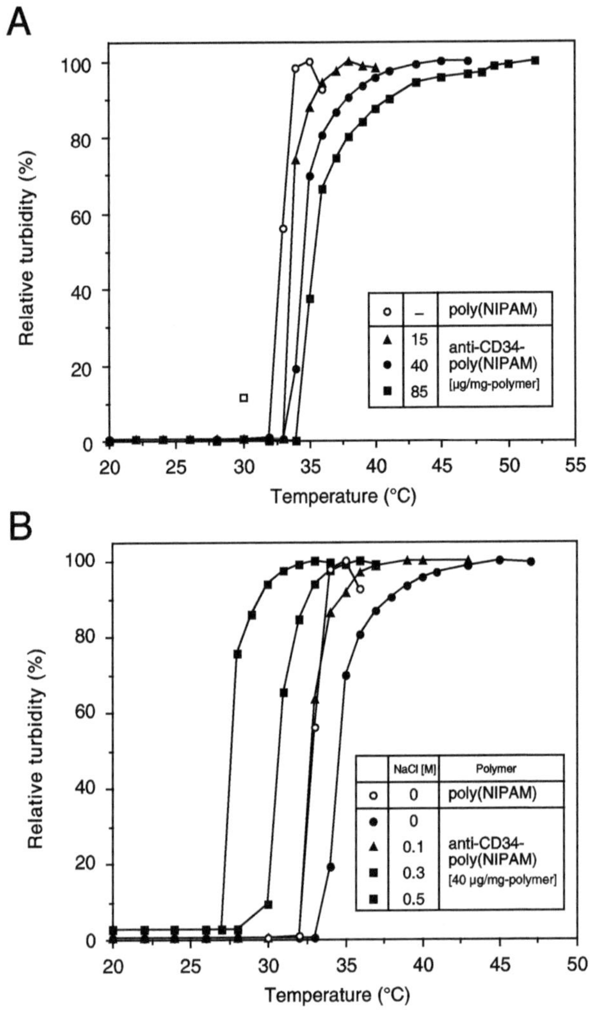

Fig. 8 Thermoprecipitation of antibody–polyNIPAM conjugates. **A** effect of degree of polyNIPAM modification (0.15 M NaCl), **B** effect of NaCl concentration

The cloud-point was increased to 37–47 °C depending on the degree of modification of the polymer and salt concentration in the solution. In the presence of more than 0.15 M NaCl, polyNIPAM-modified antibodies precipitated from the solution at or below the cloud-point temperature of native polyNIPAM. Thus, polyNIPAM-modified antibodies can be completely recovered from the solution by a change in ambient temperature. In addition, polyNIPAM and polyNIPAM-modified antibodies were completely partitioned to the upper phase when they were introduced into PEG-dextran aqueous two-phase systems (Fig. 9).

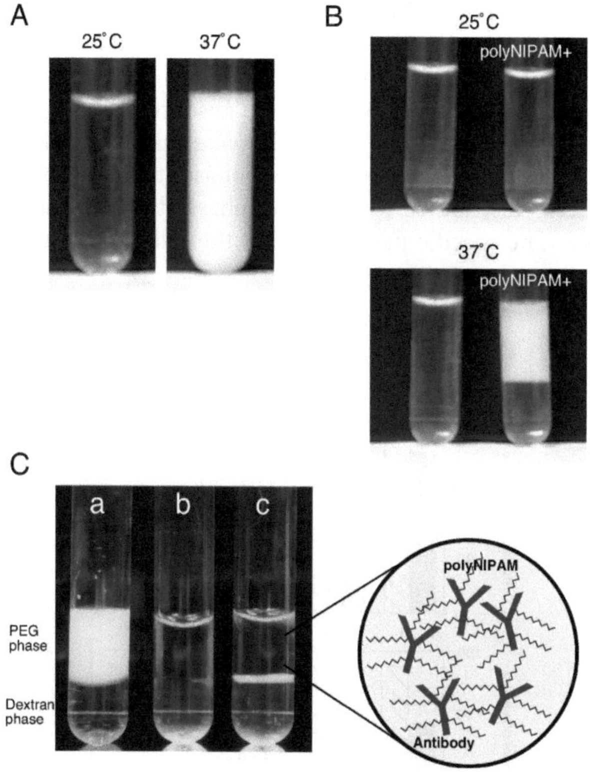

Fig. 9 Photographs of polyNIPAM solution and aqueous two-phase system containing polyNIPAM and antibody–polyNIPAM conjugates. **A** Precipitation of polyNIPAM in response to temperature change. **B** Partitioning of polyNIPAM in an aqueous two-phase system (4% PEG8000–5% dextran T500). **C** Partitioning of cells in an aqueous two-phase system (4% PEG8000–5% dextran T500) containing antibody–polyNIPAM conjugate. (*a*) Partitioning of antiCD34–polyNIPAM conjugate (32 °C). (*b*) Partitioning of KG-1 cells in the absence of antiCD34–polyNIPAM conjugate (*bottom phase*). (*c*) Partitioning of KG-1 cells in the presence of antiCD34–polyNIPAM conjugate after centrifugation at 1000 rpm for 5 min (interface)

4.3 Distribution Behavior of Cells

In a previous study on the partitioning of specific cells using aqueous two-phase systems, no toxic effect on animal cells was observed in an aqueous two-phase system composed of 4% PEG8000–5% dextran T500. Therefore, the system was employed for cell separation using polyNIPAM-modified antibodies. Since the distribution of cells is affected by the concentration of NaCl added to the system, the effects of the NaCl concentration on the partitioning of cells with or without polyNIPAM-modified antibodies were examined. In this experiment, three kinds of antibodies (anti-CD34, CD2 and erbB2) were modified with polyNIPAM, and the distribution of the cells expressing the surface proteins recognized by the respective antibodies, was examined (Fig. 10). Without application of the antibody conjugates, the cells preferentially distributed to the upper phase in the absence of NaCl. However, when NaCl was added at up to 0.15 M, the cells were pushed down to the bottom phase. No cell was partitioned to the upper phase at 0.15 M NaCl, whereas less than 10% of cells distributed to the upper phase at the critical concentration of 0.1 M NaCl. The specific partitioning of the cells to the upper phase was

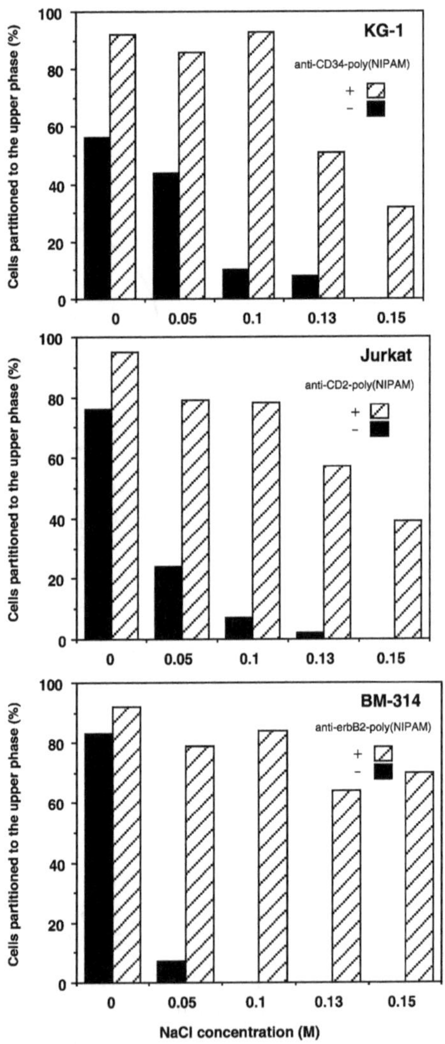

Fig. 10 Affinity partitioning of cells in an aqueous two-phase system (4% PEG8000–5% dextran T500) containing antibody–polyNIPAM conjugates

accomplished by adding the antibody conjugates in the aqueous two-phase system. Ideally, it is preferable that the cells bound to the antibody conjugates are partitioned to the upper phase and all of the other cells are partitioned to the bottom phase. When the NaCl concentration increased, the distribution of the specific cells to the bottom phase increased. Since the antibody conjugates were always partitioned to the upper phase without applying the cells and since the cells tended to partition to the bottom phase in the presence of a high concentration of NaCl, the cells seem to be pushed down to the bottom phase when an insufficient amount of the antibody conjugates bound to the cells. In the all antibody-cell combination, the ratio of specific partitioning of the cells to the nonspecific partitioning became maximal on the addition of 0.1 M NaCl. Therefore, this was determined as the optimal condition for specific cell separation using the antibody conjugates in the aqueous two-phase system. Figure 11 shows cell partitioning at different ligand densities. The increase in the ligand density is effective in improving the specific partitioning of cells to the upper phase. Maximum cell partitioning (93%) was observed when the ligand concentration was 40 μg/mg polymer (80 μg/1.4 g-system) for the treatment of 3.5×10^5 cells. Table 3 shows the partitioning coefficient (ratio of the cell concentration distributed to the upper phase to the cell concentration distributed to the bottom phase) of the cells. Almost all of the cells partitioned to the bottom phase without the additive or with native polyNIPAM. With the addition of the antibody conjugates, more than 90% of cells distributed to the upper phase. Since the antibodies modified with polyNIPAM can be precipitated from the solution by temperature change, the unused antibodies were recovered from the upper phase and reused for the second cycle of cell separation. Although the partition coefficient of cells for

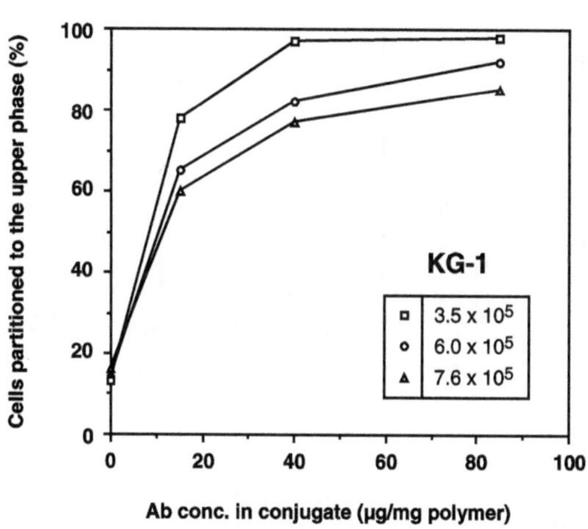

Fig. 11 Partitioning of cells at different concentrations of antibody–polyNIPAM conjugate in the aqueous two-phase system

Table 3 Effect of antibody conjugate on partition coefficient of cells

	Partition coefficient (–) KG-1	Jurkat	BM-314
Control (without conjugate)	0.12 (10%)	0.08 (7%)	0 (0%)
PolyNIPAM	0.19 (15%)	0.19 (15%)	0.08 (7%)
Antibody–polyNIPAM conjugate	18.6 (93%)	8.7 (78%)	7.0 (84%)
Recycled antibody–polyNIPAM conjugate	5.0 (75%)	3.9 (70%)	3.5 (73%)

Values in parentheses are the percentage of viable cells partitioned to the upper phase

the second separation decreased due to the decreased antibody conjugates added to the two-phase system by the incomplete recovery of the antibodies from the upper phase, it is possible to use the recovered antibody conjugates for cell separation (Table 3).

4.4
Specific Separation from Cell Mixture

As a model cell separation system, the specific separation of KG-1 cells from Jurkat cells was examined using anti-CD34 antibodies modified with polyNI-

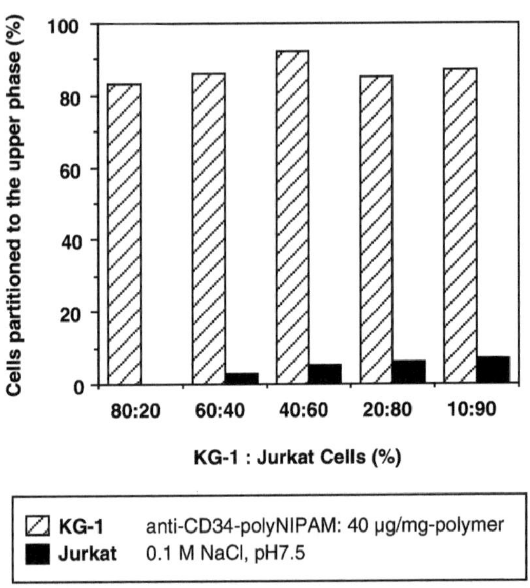

Fig. 12 Specific separation of the mixture of KG-1 and Jurkat cells

PAM in the aqueous two-phase system (Fig. 12). CD34 surface antigen is recognized as an important marker for hematopoietic stem cells and some somatic stem cells. Thus, this system can be a good model for the separation of CD34-positive cells from bone marrow or peripheral blood. More than 80% of the KG-1 cells (CD34-positive) were partitioned to the upper phase for the all cell ratio tested. Complete separation was achieved when the cell ratio was 80 (KG-1) : 20 (Jurkat). As the Jurkat cell proportion was higher in the cell mixture, the KG-1 cells in the upper phase were slightly contaminated with Jurkat cells. It was possible to achieve a more complete separation of KG-1 from Jurkat cells when the separated cell mixture was again applied to the aqueous two-phase system containing the antibody conjugates or when the NaCl concentration was increased to 0.15 M, although the recovery of cells slightly decreased. In the practical separation of stem cells, target cells often occupied a small proportion of the cell mixture (e.g., 1–4% $CD34^+$ cells in human bone marrow cells). Therefore, it is necessary to design a separation procedure accompanied by recovery and purity according to the purpose.

5 Conclusion

In the field of regenerative medicine, it seems that the demand for the separation of functional cells is increasing. Now, it is thought that cell sorting using a flow cytometer is the most powerful and decisive method for cell separation, but it is still necessary to provide a separation method that satisfies the demand for each separation. The cell separation method with an aqueous two-phase system using antibody conjugates described here, will be very useful, when large numbers of cells are to be processed for the separation of specific cells based on the antibody binding to a cell-surface marker protein. In order to separate specific cells precisely, the use of monoclonal antibodies is essential. Many marker proteins are identified for various cell types, and many antibodies against marker proteins are commercially available. However, since such antibodies are very expensive, the efficient use of antibodies is necessary in large scale processing. The method may have an advantage in this point.

Judging from the results that specific cell portioning to the upper phase decreased in the presence of higher concentrations of NaCl, it is assumed that the cells partitioned to the upper phase at higher concentrations of NaCl are expressing the marker protein at a high level. In the analysis using a flow cytometer, it is possible to separate cells based on the difference in the expression level and the combination of plural marker proteins. With the present method, by changing the salt concentration in the system and/or polymer composition to be modified to antibodies, a more precise separation based on cell-surface marker proteins may become possible.

The separation process of specific cells will become more and more important in the realization of higher medical care using the stem cell technology. It is still necessary to develop effective and reliable separation procedures for specific cells.

References

1. Thomson JA, Itskovitz-Eldor J, Shapiro SS, Waknitz MA, Swiergiel JJ, Marshall VS, Jones JM (1998) Science 282:1145
2. Reubinoff BE, Pera MF, Fong CY, Trounson A, Bongso A (2000) Nat Biotechnol 18:399
3. Wobus AM, Boheler KR (2005) Physiol Rev 85:635
4. Bongso A, Richards M (2004) Best Pract Res Clin Obstet Gynaecol 18:827
5. Stojkovic M, Lako M, Strachan T, Murdoch A (2004) Reproduction 128:259
6. Barry FP, Murphy JM (2004) Int J Biochem Cell Biol 36:568
7. Okano H, Yoshizaki T, Shimazaki T, Sawamoto K (2002) Parkinsonism Relat Disord 9:23
8. Rahaman MN, Mao JJ (2005) Biotechnol Bioeng 91:261
9. Conrad C, Huss R (2005) J Surg Res 124:201
10. Yamada S, Kojima I (2005) J Hepatobiliary Pancreat Surg 12:218
11. Baizabal JM, Furlan-Magaril M, Santa-Olalla J, Covarrubias L (2003) Arch Med Res 34:572
12. Gunsilius E, Gastl G, Petzer AL (2001) Biomed Pharmacother 55:186
13. Morshead CM (2004) Dev Neurosci 26:93
14. Gimble JM, Guilak F (2003) Cytotherapy 5:362
15. Hristov M, Erl W, Weber PC (2003) Trends Cardiovasc Med 13:201
16. Okano H (2002) J Neurosci Res 69:698
17. Wognum AW, Eaves AC, Thomas TE (2003) Arch Med Res 34:461
18. Zipori D (2004) Blood Cells Mol Dis 33:211
19. Kirchhof N, Harder F, Petrovic S, Kreutzfeldt S, Schmittwolf C, Durr M, Kirsten J, Muhl B, Merkel A, Muller AM (2002) Cells Tissues Organs 171:77
20. Weissman IL, Anderson DJ, Gage F (2001) Annu Rev Cell Dev Biol 17:387
21. Sharpe PT (1988) Methods of cell separation. Elsevier Science Publishers, Amsterdam
22. Kumar RK, Lykke AW (1984) Pathology 16:53
23. Bauer J (1999) J Chromatogr B Biomed Sci Appl 722:55
24. Griffith OM (1978) Anal Biochem 87:97
25. Pertoft H (2000) J Biochem Biophys Methods 44:1
26. Herzenberg LA, Parks D, Sahaf B, Perez O, Roederer M, Herzenberg LA (2002) Clin Chem 48:1819
27. Tanke HJ, van der Keur M (1993) Trends Biotechnol 11:55
28. Orfao A, Ruiz-Arguelles A (1996) Clin Biochem 29:5
29. Herzenberg LA, De Rosa SC (2000) Immunol Today 21:383
30. Ashcroft RG, Lopez PA (2000) J Immunol Methods 243:13
31. Ibrahim SF, van den Engh G (2003) Curr Opin Biotechnol 14:5
32. So PL, Epstein EH Jr (2004) Trends Biotechnol 22:493
33. Molday RS, Molday LL (1984) FEBS Lett 170:232
34. Vartdal F, Gaudernack G, Funderud S, Bratlie A, Lea T, Ugelstad J, Thorsby E (1986) Tissue Antigens 28:301

35. Hardwick RA, Kulcinski D, Mansour V, Ishizawa L, Law P, Gee AP (1992) J Hematother 1:379
36. Kemshead JT (1992) J Hematother 1:35
37. Ugelstad J, Stenstad P, Kilaas L, Prestvik WS, Herje R, Berge A, Hornes E (1993) Blood Purif 11:349
38. Safarik I, Safarikova M (1999) J Chromatogr B Biomed Sci Appl 722:33
39. Gupta AK, Gupta M (2005) Biomaterials 26:3995
40. Kumar A, Kamihira M, Galaev IY, Mattiasson B (2001) Biotechnol Bioeng 75:570
41. Albertsson PA (1986) Partition of Cell Particles and Macromolecules. 3rd edn. Wiley, New York
42. Walter H, Brooks DE, Fisher D (1985) (eds) Partitioning in aqueous two-phase systems: theory, methods, uses, and applications to biotechnology. Academic Press, New York
43. Hatti-Kaul R (ed) (2000) Aqueous two-phase systems: methods and protocols. Humana Press, Totowa, New Jersey
44. Walter H, Fisher D (1985) Separation and subfractionation of selected mammalian cell populations. In: Walter H, Brooks DE, Fisher D (eds) Partitioning in aqueous two-phase systems: theory, methods, uses, and applications to biotechnology. Academic Press, New York, p 378
45. Hamamoto R, Kamihira M, Iijima S (1996) J Ferment Bioeng 82:73
46. Zijlstra GM, Michielsen M, De Gooijer CD, van der Pol LA, Tramper J (1996) Biotechnol Prog 12:363
47. Goubran-Botros H, Birkenmeier G, Otto A, Kopperschlager G, Vijayalakshmi MA (1991) Biochim Biophys Acta 1074:69
48. Laboureau E, Capiod JC, Dessaint C, Prin L, Vijayalakshmi MA (1996) J Chromatogr B 680:189
49. Shanbagh VP, Axelsson CG (1975) Eur J Biochem 60:17
50. Harris JM (1985) J Macromol Sci Rev Polym Chem Phys C-25:325
51. Kopperschlger G, Johansson G (1982) Anal Biochem 124:117
52. Johansson G, Andersson M (1984) J Chromatgr 303:39
53. Wuenschell GE, Naranjo E, Arnold FH (1990) Bioprocess Eng 5:199
54. Franco TT, Galaev IY, Hatti-Kaul R, Niklas H, Blow L, Mattiasson B (1997) Biotechnol Techniques 11:231
55. Pietruszka N, Galaev IY, Kumar A, Brzozowski ZK, Mattiasson B (2000) Biotechnol Prog 16:408
56. Sharp KA, Yalpani M, Howard SJ, Brooks DE (1986) Anal Chem 154:110
57. Karr LJ, Shafer SG, Harris JM, Van Alstine JM, Snyder RS (1986) J Chromatogr 354:269
58. Hughes P, Lowe CR (1988) Enzyme Microb Technol 10:115
59. Kamihira M, Kaul R, Mattiasson B (1992) Biotechnol Bioeng 40:1381
60. Suzuki M, Kamihira M, Shiraishi T, Takenchi H, Kobayashi T (1995) J Ferment Bioeng 80:78
61. Johansson HO, Svensson M, Persson J, Tjerneld F (2000) Aqueous two-phase systems with smart polymers. In: Galaev IY, Mttiasson B (eds) Smart polymers for bioseparation and bioprocessing. Harwood Academic Publishers, Chur, Switzerland, p 78
62. Schild HG (1992) Prog Polym Sci 17:163
63. Takei YG, Aoki T, Sanui K, Ogata N, Okano T, Sakurai Y (1993) Bioconjug Chem 4:341
64. Kumar A, Galaev IY, Mattiasson B (1998) Biotechnol Bioeng 59:695

Author Index Volumes 101–108

Author Index Volumes 1–50 see Volume 50
Author Index Volumes 51–100 see Volume 100

Subject Index

MIX
Papier aus verantwortungsvollen Quellen
Paper from responsible sources
FSC® C105338

If you have any concerns about our products,
you can contact us on
ProductSafety@springernature.com

In case Publisher is established outside the EU,
the EU authorized representative is:
Springer Nature Customer Service Center GmbH
Europaplatz 3, 69115 Heidelberg, Germany

Printed by Libri Plureos GmbH
in Hamburg, Germany